The Natural Contract

Studies in Literature and Science
published in association with the
Society for Literature and Science

Titles in the series

Transgressive Readings: The Texts of Franz Kafka and Max Planck
by Valerie D. Greenberg

A Blessed Rage for Order: Deconstruction, Evolution, and Chaos
by Alexander J. Argyros

Of Two Minds: Hypertext Pedagogy and Poetics by Michael Joyce

The Artificial Paradise: Science Fiction and American Reality
by Sharona Ben-Tov

Conversations on Science, Culture, and Time
by Michel Serres with Bruno Latour

Genesis by Michel Serres

The Natural Contract by Michel Serres

MICHEL SERRES

The Natural Contract

Translated by
Elizabeth MacArthur and William Paulson

Ann Arbor

THE UNIVERSITY OF MICHIGAN PRESS

Published in the United States of America by
The University of Michigan Press
Manufactured in the United States of America
♾ Printed on acid-free paper
1998 1997 1996 1995 4 3 2 1

A CIP catalogue record for this book is available from the British Library.

Library of Congress Cataloging-in-Publication Data

Serres, Michel.
[Contrat naturel. English]
The natural contract / Michel Serres ; translated by Elizabeth MacArthur and William Paulson.
p. cm.
ISBN 0-472-09549-8 (alk. paper). — ISBN 0-472-06549-1 (pbk. : alk. paper)
1. Environmental sciences—Philosophy. 2. Environmental responsibility. I. Title.
GE60.S4713 1995
363.7—dc20 95-2685
CIP

The publisher is grateful for partial subvention for translation from the French Ministry of Culture.

For Robert Harrison,

... *casu quodam in silvis natus* ...

(Livy I, 3)

Translators' Acknowledgments

It has been both a signal pleasure and a daunting task to translate the writing of Michel Serres, himself a consummate translator of ideas from one idiom to another. We have tried to make our English version clear and fluent, while still preserving something of his inimitable style, word play, and breadth of meaning. The range of domains to which Serres refers, often simultaneously, poses particular challenges to the translator, and we found ourselves consulting sailors and classicists, lawyers and mathematicians. We would like to express our gratitude to the following people who gave us advice: Robert Bourque, H. D. Cameron, Stephanie Castleman, Hervé Pisani, Jacqueline Simons, Stephen Simons, Katherine Staton. We also read with profit Felicia McCarren's translation of chapter 2, "Natural Contract," which appeared in *Critical Inquiry* 19 (Autumn 1992): 1–21.

Above all we would like to thank Michel Serres for his generous help in a number of lengthy faxes and a sunny conversation in Santa Barbara. His cooperation enabled us to avoid several misreadings and to clarify in English some difficult passages in the original French. We take full responsibility for the misreadings and infelicities that remain.

Contents

War, Peace

A pair of enemies brandishing sticks is fighting in the midst of a patch of quicksand. Attentive to the other's tactics, each answers blow for blow, counterattacking and dodging. Outside the painting's frame, we spectators observe the symmetry of their gestures over time: what a magnificent spectacle—and how banal!

The painter, Goya, has plunged the duelists knee-deep in the mud. With every move they make, a slimy hole swallows them up, so that they are gradually burying themselves together. How quickly depends on how aggressive they are: the more heated the struggle, the more violent their movements become and the faster they sink in. The belligerents don't notice the abyss they're rushing into; from outside, however, we see it clearly.

Who will die? we ask. Who will win? they are wondering—and that's the usual question. Let's make a wager. You put your stakes on the right; we've bet on the left. The fight's outcome is in doubt simply because there are two combatants, and once one of them wins there will be no more uncertainty. But we can identify a third position, outside their squabble: the marsh into which the struggle is sinking.

For here the bettors are in the same doubt as the duelists, and both bettors and duelists are at risk of losing collectively, since it is more than likely that the earth will swallow up the fighters before they and the gamblers have had a chance to settle accounts.

On the one hand there's the pugnacious subject, every man for himself; on the other, the bond of combat, so heated that it inflames the audience, enthralled to the point of joining in with its cries and coins.

But aren't we forgetting the world of things themselves, the sand, the water, the mud, the reeds of the marsh? In what quicksands are we, active adversaries and sick voyeurs, floundering side by side? And I who write this, in the solitary peace of dawn?

Achilles, king of war, struggles against a swelling river. Strange, mad battle! We don't know if Homer, in book 21 of the *Iliad,* takes this river to be the mounting tide of furious enemies who assail the hero.

In any case, as he throws the innumerable corpses of adversaries vanquished and killed into the current, the level rises so that the stream, bursting its banks, reaches up to his shoulders to threaten him. Then, shaken by a new terror, he casts off bow and saber; his free hands raised toward the heavens, he prays. Is his triumph so total that his repugnant victory is transformed into defeat? In place of his rivals the world and the gods burst into view.

History, dazzling in its truth, unveils the glory of Achilles or some other hero, whose valor comes from laurels won in limitless, endlessly renewed war. Violence, with its morbid luster, glorifies the victors for propelling the motor of history. Woe to the vanquished!

A first step toward humanization came from proclaiming the victims of this animal barbarity more blessed than the murderers.

As a second step, now, what is to be done with this river, once mute, which is starting to burst its banks? Does the swelling come from the springtime or from the squabble? Must we distinguish two battles: the historical war waged by Achilles against his enemies and the blind violence done to the river? A new flood: the level is rising. Fortunately, on that day, during the Trojan War, fire from the heavens dried up the waters; unfortunately, without promising any alliance.

River, fire, and mud are reminding us of their presence.

Nothing ever interests us but spilled blood, the manhunt, crime stories, the point at which politics turns into murder; we are enthralled only by the corpses of the battlefield, the power and glory of those who hunger for victory and thirst to humiliate the losers; thus entertainment mongers show us only corpses, the vile work of death that founds and traverses history, from the *Iliad* to Goya and from academic art to prime-time television.

Modernity, I notice, is beginning to tire of this loathsome culture. In the present era, murderous winners are admired somewhat less, and despite the glee with which killing fields are put on display, they draw only unenthusiastic applause: these are, I presume, good tidings.

In these spectacles, which we hope are now a thing of the past, the adversaries most often fight to the death in an abstract space, where they struggle alone, without marsh or river. Take away the world around the battles, keep only conflicts or debates, thick with humanity and purified of things, and you obtain stage theater, most of our narratives and philosophies, history, and all of social science: the interesting spectacle they call cultural. Does anyone ever say *where* the master and slave fight it out?

Our culture abhors the world.

Yet quicksand is swallowing the duelists; the river is threatening the fighter: earth, waters, and climate, the mute world, the voiceless things once placed as a decor surrounding the usual spectacles, all those things that never interested anyone, from now on thrust themselves brutally and without warning into our schemes and maneuvers. They burst in on our culture, which had never formed anything but a local, vague, and cosmetic idea of them: nature.

What was once local—this river, that swamp—is now global: Planet Earth.

Climate

Let us propose two equally plausible interpretations of the stable high-pressure zones over North America and Europe in 1988 and 1989.

The first interpretation: a similar sequence of hot dry days could easily be found in the decades for which we have records, or inferred for the millennia beyond human memory. The climatic system varies greatly, and yet fairly little, being relatively invariant in its variations: quick and slow, catastrophic and mild, regular and chaotic. Rare phenomena are therefore striking, but they shouldn't surprise us.

Some stone blocks that hadn't moved since the gigantic flows of the receding Ice Age, at the end of the Quaternary, came down in 1957, carried along by the extraordinary flooding of the Guil, an ordinary Alpine torrent. When will they move a third time? Next year or in twenty thousand. There's nothing unnatural about this uncertainty; that's just the way it is.

The rarest of events can be integrated (or acclimated, as they say) into meteorology, where the irregular becomes all but normal. A summery winter fits into the pattern: nothing to write home about.

Yet meanwhile the concentration of carbon dioxide has been growing in the atmosphere since the industrial revolution, a byproduct of fossil fuels; the propagation of toxic substances and acidifying products is increasing; the presence of other greenhouse gases is growing. The sun warms the earth, which in turn radiates part of that heat back out into space; an overly thick dome of carbon dioxide would allow the sun's radiation to pass through but would trap the heat radiating back; normal cooling would then slow down, and evaporation would be modified, just as in a greenhouse. So is the earth's atmosphere in danger of becoming more like that of Venus, unlivable?

The past, however distant, never knew such experiences. Because of our actions, the composition of the air, and thus its physical and chemical properties, is changing. Is the behavior of the system suddenly going to be disrupted? Is it possible to describe, estimate, calculate, even conceive, and ultimately steer this global change? Will the climate become warmer? Can one foresee some of the consequences of such transformations and expect, for example, a sudden or gradual rise in sea levels? What would become, then, of all the low countries—Holland, Bangladesh, or Louisiana—submerged beneath a new deluge?

According to the second interpretation, this is something new under the sun, something rare and abnormal, whose causes can be evaluated but whose consequences cannot: can it be acclimated by standard climatology?

At stake is the Earth in its totality, and humanity, collectively.

Global history enters nature; global nature enters history: this is something utterly new in philosophy.

Does the stable sequence of hot and dry days that Europe recently enjoyed or worried about point to man-made acts rather than to variables considered natural? Will the floods come from springtime or from an attack? We surely don't know; what is more, all our knowledge, with its hard-to-interpret models, contributes to this uncertainty.

Thus in doubt, will we refrain from taking action? That would be imprudent, for we are embarked on an irreversible economic, scientific, and technological adventure; one can regret the fact, and even do so with skill and profundity, but that's how it is, and it depends less on us than on what we have inherited from history.

Wager

We must anticipate and decide. Wager, therefore, since our models can serve to defend the two opposing theses. If we judge our actions innocent and we win, we win nothing, history goes on as before, but if we lose, we lose everything, being unprepared for some possible catastrophe. Suppose that, inversely, we choose to consider ourselves responsible: if we lose, we lose nothing, but if we win, we win everything, by remaining the actors of history. Nothing or loss on one side, win or nothing on the other: no doubt as to which is the better choice.

Now this classic argument is valid when an individual subject chooses, for himself, his actions, his life, his fate, his last ends; it is conclusive, to be sure, but has no immediate application, when the subject who must decide unites more than the nations: humanity. Suddenly a local object, nature, on which a merely partial subject could act, becomes a global objective, Planet Earth, on which a new, total subject, humanity, is toiling away. These totalizations of both subject and object will require more work than was called for by the decisive argument of the wager.

But the recent conferences on the environment in Toronto, Paris, London, and The Hague testify to an anxiety that is beginning to spread. It suddenly resembles a general mobilization! More than twenty-five countries have recently signed an agreement for the common governance of the problem. The crowd is massing like clouds before the storm, which may or may not break, no one knows. Old-style groups are working together on a new globality, which is starting to coalesce in the same way that nature seems to come together as a whole in the best scientific works.

Air raid warning! Not a danger coming in from space, but the risk run on earth by the atmosphere: by the weather or climate understood as global systems and as general conditions of survival. For the first time, could the West—which hates children, since it produces so few and doesn't want to pay for the education of those remaining—be starting to think about its descendants' breathing? Long confined to the short term, could the West now make a long-term projection? Could science, overwhelmingly analytic, consider an object in its totality for the first time? In the face of the threat, could notions or scientific disciplines unite, like the nations? Are our thoughts, until recently rooted exclusively in their own history, rediscovering geography, essential and exquisite? Could philosophy, once alone in thinking globally, be dreaming no longer?

Having thus stated in general terms the climate problem, with its indeterminacy, we can discover its immediate causes, but we can also evaluate its deep and remote conditions, and finally seek possible solutions to it. In the economy, in industry, in all of technology, and in demography lie immediate reasons with which we are all familiar, though we are not able to act on them easily. We must also fear that the short-term solutions proposed by these disciplines would reproduce the causes of the problem by reinforcing them.

The long-term causes are less obvious; they must now be set forth.

War

General mobilization! I purposely use the term employed at the beginning of wars. Air raid warning! I deliberately use the alert given in land or sea combat.

Suppose, then, a battle situation. Schematically, it sets up two adversaries, alone or numerous, each side either armed or not with weapons that are either more or less powerful, duelists equipped with sticks, heroes armed with sabers and bows. When the engagement is over, the day's or campaign's outcome entails not only decisive victory and defeat but losses to be deplored: deaths and destruction.

Suppose that these losses increase rapidly, in obvious proportion to the energy of the means that are mobilized. The most extreme known case was the recent situation in which we couldn't decide whether or not the nuclear arsenal, through anticipation of the damages that would be inflicted and shared by the belligerents, was guaranteeing the more or less stable peace that the nations that had set up the arsenal experienced for forty years. Although we weren't sure, we suspected that that was the case.

To my knowledge it has never been remarked that this growth would overturn the initial schema once it reached a certain globality. At the outset we posited two rivals facing off, as in Goya's quicksand, to decide once and for all on a loser and a winner. Perhaps because of a threshold effect, the sharing of destruction and the increase in its means produce an astonishing reversal: suddenly, the two enemies find themselves in the same camp, and, far from giving battle to one another, they struggle together against a common third competitor. Which one?

It is hidden by the heatedness of the conflict and the often tragic magnitude of the human stakes involved. The duelists don't see that they're sinking into the muck, nor the warriors that they're drowning in the river, together.

In its burning heat, history remains blind to nature.

Dialogue

Let's examine an analogous situation. Suppose two speakers, determined to contradict each other. As violent as their confrontation may be, as long as they are willing to continue the discussion they must speak a common language in order for the dialogue to take place. There can't be an argument between two people if one speaks a language the other can't understand.

To shut someone else up, somebody suddenly changes idiom: thus doctors once spoke Latin, and collaborators during the last war German, just as today's Parisian newspapers use English, so that the good people, understanding nothing, obey in a daze. Nearly all technical words are harmful in science and philosophy;

they serve only to separate the sectarians of the parish from those who are excluded from the conversation so that the masters can hold on to some form of power.

Even more than a common language, debate requires the speakers to use the same words in a sense that is at least related and at best identical. They therefore enter into a preliminary contract, spoken or unspoken, stipulating the use of a common code. This agreement, most often tacit, precedes the debate or combat, which, in turn, presupposes an agreement; I think that's what is signified by the term "declaration of war," whose wording allows no ambiguity: a legal contract that precedes the violent explosions of conflicts.

By definition, war is a legal state.

Furthermore, there can be no verbal squabble if a gigantic noise, coming from a new source, covers up every voice with its static. The usual procedure in battles of airwaves and images: jamming. In the evening, at home, television's clamor silences any discussion. An old phonograph ad—"His Master's Voice"—shows a well-behaved dog sitting with ears pricked up in front of a gramophone horn; we have become obedient puppies, passively listening to our masters' uproar. We never talk any more, that's for sure. To keep us from it, our civilization sets motors and loudspeakers screaming.

And we no longer remember that the (now quite rare) word *noise,* used (in French) only in the sense of quarrel—in the expression *chercher noise* (to pick a quarrel)—that this word, in the Old French from which it comes, meant tumult and furor. English took from us the sense of *sound* while we kept that of *battle.* Still further back, in the original Latin, the heaving of water could be heard, the roaring and lapping. *Nauticus:* navy, nausea (do we get seasick from hearing?), noise.

In short, the two opponents in a dialogue struggle together, on the same side, against the noise that could jam their voices and their arguments. Listen to them raise their voices, concertedly, when the brouhaha begins. Debate, once again, presupposes this agreement. The quarrel, or *noise* in the sense of battle, supposes a common battle against the jamming, or *noise* in the sense of sound.

With this, the initial schema is complete: two speakers whom we see clearly are tenaciously contradicting each other, but there, present, two invisible if not tacit specters are keeping a vigil. The first specter is a mutual friend who conciliates the speakers by the (at least virtual) contract of common language and defined words; the second specter is a mutual enemy against whom they actually struggle, with all their combined forces: this noisy noise, this jamming, which would cover up their own din to the point of nullifying it. To exist, war must make war on that war. And no one notices this.

So in the end every dialogue is like a game for four players, played on a new figure, a square or a cross. The two parties to the dispute exchange fair arguments or low-down insults along one diagonal, while on the second, sideways or across, most often without the speakers' knowledge, their contractual language fights inch by inch against the ambient noise to preserve its purity.

In the first case it's a subjective battle, that is, one between subjects, the adversaries; but in the second case it's an objective battle, between two nameless agencies that as yet have no legal status, because the phenomenal spectacle of the noisy and inflamed dialogue still hides them and distracts our attention.

The debate hides the true enemy.

The adversaries no longer exchange words but rather, without saying a thing, blows. Someone fights someone else, subject face-to-face with subject. Soon, because fists are no longer enough for their rage, the two adversaries gather stones, refine them, invent iron, swords, armor, and shields, discover gunpowder, then put it to use, find thousands of allies, assemble in giant armies, multiply their battlefronts, at sea, on land, and in the air, seize the power of atoms and take it to the stars—is there anything simpler and more monotonous than this history? We need to take stock of the situation again now that we've reached the end point of growth.

Never mind the millions of deaths: as soon as war was declared, the belligerents understood that blood and tears would flow, and had accepted the risk. The outcome was almost voluntary; there was nothing unexpected about it. Does there exist in this carnage a threshold of the intolerable? Our histories don't indicate that.

Never mind the so-called material losses either: ships, tanks and cannons, aircraft, equipment, transports, and cities, all annihilated. This destruction, too, is accepted from the moment the belligerents open hostilities, using weapons constructed by human hands, which the enemies, if I may say so, have at hand.

But we never speak of the damage inflicted on the world itself by these wars, once the number of soldiers and the means of fighting grow in strength. With the declaration of war, the belligerents do not consciously accept this damage, but in reality they produce it together, out of the objective fact of belligerence. They tolerate it unawares. There's no clear consciousness of the risks incurred, except, sometimes, by the wretched, the third parties excluded from noble struggles: that picture of the field of oats devastated by the knightly battle, we don't remember anymore if we saw it as an illustration in old history texts or in those books to which the schools of the past gave the marvelous name, "object lessons."

So now we have a fleet of sunken oil tankers, several gutted atomic submarines, a few exploded thermonuclear bombs: the subjective victory in the subjective war of so-and-so against so-and-so suddenly counts very little in the face of the objective results of the objective violence unleashed against the world by the means at the belligerents' disposal. Especially now that the objective war's outcome has global consequences.

Does the contemporary retreat before a worldwide conflict come from the fact that from now on what is at stake is things rather than people? and the global rather than the local? Is history stopping in the face of nature? At any rate that's how the Earth became the common enemy.

Until now our management of the world has been carried out through belligerence, just as historical time has been driven by struggle. A global change is underway: ours.

War and Violence

From now on, then, I will call *subjective* wars those, whether nuclear or conventional, that nations or states fight with the aim of temporary dominance—a dominance that we are skeptical about since we have noticed that those who lost the last war, and thus were disarmed, today dominate the universe. I will call *objective* violence that in which all the enemies, unconsciously joined together, are in opposition to the objective world, which is called, in an astonishing metaphor, the "theater" of hostilities. Thus the real is reduced to a spectacle in which the debate stands out against a cardboard backdrop that can be displayed or dismantled at will. For the subjective wars, things didn't exist in themselves.

And since it is customarily said of these squabbles that they are the motor of history, once again we reach the conclusion that culture abhors the world.

Now if war, or armed conflict, declared consciously, voluntarily, and according to the rules, remains a legal relationship, objective violence comes to blows without any preliminary contract.

This leads to a new square, based on the one sketched in my discussion of dialogue: the rivals of the day are on two opposing corners, fighting their battles along a diagonal. We see only them: since the dawn of history, they have produced all the entertainment, quarrels, and furor, the exciting arguments and tragic losses; they have provided all the spectacles and kept up the dialogues. This is the theater of dialectics, a logic of appearances, having the rigor of dialectics and the visibility of appearances.

But on a third corner of the same square is the worldwide world. Invisible, tacit, reduced to a stage set, it is the objective common enemy of the legal alliance between the de facto rivals. Together, along the other diagonal, crossing the first one, the rivals press with all their weight on objects, which bear the effects of their actions. Every battle or war ends up fighting against things or, rather, doing them violence.

And, as one might expect, the new adversary can win or lose.

In the days of the *Iliad* and of Goya, the world wasn't considered fragile; on the contrary, it was threatening, and it easily triumphed over men, over those who won battles, and over wars themselves. The quicksand sucks in the two combatants together; the stream threatens to engulf Achilles—the victor?—after having swept away the corpses of the vanquished.

The global change now underway not only brings history to the world but also makes the power of the world precarious, infinitely fragile. Once victorious, the Earth is now a victim. What painter will depict the deserts vitrified by our war games? What visionary poet will lament vile, bloody-fingered dawn?

But people are dying of hunger in the deserts just as they are suffocating in the slimy quicksand or drowning in the rising rivers. Conquered, the world is finally conquering us. Its weakness forces strength to exhaust itself and thus our own strength to become gentler.

The enemies' agreement to enter into war does violence, without prior agreement, to things themselves, which can in return do violence to their agreement. The new square, which shows the two rivals on two opposite corners, restores the presence of invisible, fearsome players in the other two corners: the worldwide world of things, the Earth; the worldly world of contracts, the law. The heat and noise of our spectacular struggles hides these players.

Better yet, consider instead the diagonal of subjective wars as the trace, in the plane of the square, of a revolving circle. As uncountable as the ocean's waves, varied but monotonous, and just as inevitable, these wars were said to constitute the motor of history, and in fact they constituted its eternal return: nothing new under the sun that Joshua stopped so that the battle could continue unabated. Identical in their perpetually recurring structure and dynamic, these wars grow in range, scale, means, and results. The pace accelerates, but in an infinite cycle.

The square turns, standing on one of its corners: such a rapid rotation that the rivals' diagonal, spectacularly visible, appears to become immobile, horizontal, invariant through the variations of history. The other diagonal of the gyroscope, forming a cross with this first one, becomes the axis of rotation, all the more immobile the faster the whole thing moves: a single objective violence, oriented more and more consistently toward the world. The axis rests and weighs on it. The more the subjective combat gains in means of destruction, the more the fury of the objective combat becomes unified and fixed.

A limit is reached: a certain history comes to an end when the efficacy of objective violence, which is tragic in a new way, and involuntary, replaces the useless vanity of subjective wars, which increase their arms and multiply their devastation in longing pursuit of decisive victory. These wars must be taken up again at ever closer intervals, so much does the duration of empires dwindle.

Dialectics can be reduced to the eternal return, and the eternal return of wars brings us to the world. What has for several centuries been called history is reaching this limit point, this frontier, this global change.

Law and History

War must be defined as one of the legal relations between groups or nations: a de facto state, to be sure, but above all a legal, de jure state. Since the archaic times of the first Roman laws, and doubtless even earlier, war has lasted only from the very precise procedures of the declaration until those of the armistice, duly signed by those in charge, one of whose principal prerogatives is precisely the power to decide on the opening and the cessation of hostilities. War is characterized not by the brute explosion of violence but by its organization and its legal status. And, as a result, by a contract: two groups decide, by a common agreement on which they give rulings, to devote themselves to battles, pitched or otherwise. We find once more, conscious if not written, the tacit contract between the debaters of a moment ago.

History begins with war, understood as the closure and stabilization of violent engagement within juridical decisions. The social contract that gave birth to us is perhaps born with war, which presupposes a prior agreement that merges with the social contract.

Before or beside this contract, in the otherwise limitless unleashing of pure and de facto violence, foundational and without end, groups constantly ran the risk of extinction, because vengeance begets vengeance and never stops. The cultures that did not invent these procedures for limiting the duration of violence have been erased from the face of the earth and can no longer testify to this danger. Did they even exist? It is as if in order to survive we had to pass through the filter of this war contract, which gave birth to our history by saving us from pure and thus truly deadly violence.

Violence before; war afterwards; legal contract in between.

Thus Hobbes is off by a whole era when he calls the state preceding the contract a "war of all against all," for belligerence presupposes this pact whose appearance ten philosophies attempt to explain. When everyone fights against everyone, there is no state of war, but rather violence, a pure, unbridled crisis without any possible cessation, and the participating population risks extinction. In fact and by law, war itself protects us from the unending reproduction of violence.

Jupiter, god of laws and of the sacred, obviously saves us from violence; Quirinus, god of the economy, distances us from it as well, of course; but, though it might seem paradoxical, Mars, god of war, also manages to protect us from it, even more directly than Jupiter and Quirinus: because war makes the judicial intervene at the heart of the most primitive aggressive relations. What is a conflict? Violence plus some contract. And how could this contract appear if not as a first modification of these primitive relations?

War is the motor of history: history begins with war and war set history on its course. But since, in the straitjacket of the law, war follows the repetitive dynamics of violence, the resulting movement, which always follows the same laws, mimics an eternal return. Basically we always engage in the same conflicts, and the presidential decision to release a nuclear payload imitates the act of the Roman consul or the Egyptian pharaoh. Only the means have changed.

The wars I call subjective are thus defined by the law: they begin with history and history begins with them. Juridical reason doubtless saved the local cultural subsets of which we are descendants from the automatic extinction to which those who did not invent it were condemned, without appeal, by self-perpetuating violence.

Now if there is a law, and thus a history, for subjective wars, there is none for objective violence, which is without limit or rule, and thus without history. The growth of our rational means carries us off, at a speed difficult to estimate, in the direction of the destruction of the world, which, in a rather recent backlash, could condemn us all together, and no longer by locales, to automatic extinction. Suddenly we are returning to the most ancient times, whose memory has been preserved only in and through the ideas of philosophers who theorize the law, times when our cultures, saved by a contract, invented our history, which is defined by forgetting the state that preceded it.

In conditions very different from this first state, but nonetheless parallel, we must, therefore, once again, under the threat of collective death, invent a law for objective violence. We find ourselves in the same position as our unimaginable ancestors when they invented the oldest law, which transformed their subjective violence, through a contract, into what we call wars. We must make a new pact, a new preliminary agreement with the objective enemy of the human world: the world as such. A war of everyone against everything.

If we must renew our ties with a history's foundations, that is a clear indication that we are seeing its end. Is this the death of Mars? What are we going to do with our armies? This astonishing question has come back to haunt our governments.

But more than that is at stake: the necessity to revise and even re-sign the primitive social contract. This unites us for better and for worse, along the first diagonal, without the world. Now that we know how to join forces in the face of danger, we must envisage, along the other diagonal, a new pact to sign with the world: the natural contract.

Thus the two fundamental contracts intersect.

Competition

If we move from war to economic relations, nothing notable changes in the argument. Quirinus, god of production, or Hermes, who presides over exchanges, can sometimes keep back violence more effectively than Jupiter or Mars, but they do so using the same methods as Mars. One god in several persons, then, Mars calls war what the first two call competition: the pursuit of military operations by other means—exploitation, commodities, money, or information. Even more hidden, the real conflict reappears. The same schema is renewed: by their ugliness and by the filth they accidentally spread around, chemical factories, large-scale livestock raising, nuclear reactors, and supertankers bring on objective global violence once again, with no arms other than the power of their size, no end other than the common and contractual quest for domination over men.

Let's give the name *world-object* to artifacts that have at least one global-scale dimension (such as time, space, speed, or energy): among the world-objects we know how to build, we distinguish the military ones from other purely economic or technical ones, although they produce similar results, in circumstances as rare and frequent as wars and accidents.

De facto allies for the same reasons and contracts as before, the competitors press with all their weight on the world.

We

But who is on the fourth corner of the square or at the end of the gyroscope axis? Who is doing violence to the worldwide world? What do our tacit agreements cover up? Can we draw a global picture of the worldly world, of our strictly social contracts?

On Planet Earth, henceforth, action comes not so much from man as an individual or subject, the ancient warrior hero of philosophy and old-style historical consciousness, not so much from the canonized combat of master and slave, a rare couple in quicksand, not so much from the groups analyzed by the old social sciences—assemblies, parties, nations, armies, tiny villages—no, the decisive actions are now, massively, those of enormous and dense tectonic plates of humanity.

Visible at night from orbit as the biggest galaxy of light on the globe, more populous overall than the United States, the supergiant megalopolis Europe sets out from Milan, crosses the Alps in Switzerland, follows the Rhine via Germany and the Low Countries, angles through England after crossing the North Sea, and ends up in Dublin after St. George's Channel. It's a social unit comparable to the Great Lakes or the Greenland icecap in size, in the homogeneity of its texture, and in its hold on the world. This plate of humanity has long disturbed the albedo, the circulation of water, the median temperature, and the formation of clouds or wind—in short, the elements—as well as the number and evolution of living species in, on, and under its territory.

This is the relation of man and the world today.

A major contractual actor of the human community, on the brink of the second millennium, Europe weighs at least a quarter-billion souls. Not in body weight, but in its crossed networks of relations and the number of world-objects at its disposal. It behaves like a sea.

The Earth needs only to be observed by satellite, at night, for these great dense spots to be recognized: Japan, the northeast

American megalopolis from Baltimore to Montreal, this city that is Europe, an enormous herd of monsters that Paris seems to watch over like a shepherd from afar, and the broken rim of the Dragons: Korea, Formosa, Hong Kong, Singapore. . . . When it is unevenly distributed, skyrocketing demographic growth becomes concentrated and stuck together in giant units, colossal banks of humanity as powerful as oceans, deserts, or icecaps, themselves stockpiles of ice, heat, dryness, or water; these immense units feed on themselves, advance and weigh upon the planet, for worse and for better.

Can an individual actor, lost in these gigantic masses, still say "I" when the old collectivities, themselves so lightweight, have already been reduced to uttering a paltry and outmoded "we"?

In bygone days, the individual subject was practically invisible, blended in or distributed on this Earth among the forests or mountains, the deserts and ice floes, lightweight in body and bone. There was no need for the whole universe to take up arms to crush him: a vapor, a drop of water was enough to kill him. Swallowed up like a single point, that was man of not long ago, against whom the climate was winning the war.

If we imagine that a satellite, in those eras, had been flying over the plain, what observer, on board, could have guessed at the presence of two peasants standing there at the hour of Millet's *Angelus*? Immersed in being-in-the-world, indissolubly bound with-one-another, their ploughing tools at hand, their feet plunged unto death into the immemorial soil, below the horizon, they-are-there, piously hearkening to the language of being and time, when the angel passes, the hourly bearer of the word. There's nothing more or less in our peasant or forest philosophies than in nostalgic and conventional paintings.

A frail bent reed, man thinks, knowing that he will die of this universe that, for its part, does not know that it is slaying him; he is more noble, therefore, more dignified than his conqueror because he understands this.

Nil in the universe, dissolved in the locality of being-there, man thus hadn't attained physical existence: this is his state, naturally weightless, at the hour of Millet's *Angelus* or of farm ontologies. At present he is becoming a physical variable through an exchange of power, weakness, and fragility. No longer swallowed up like a dimensionless point, he exists as a collectivity, transcending the local to extend into immense tectonic plates, just as astronomically observable as the oceans. Not only can he take up arms to crush the universe, through science and technology, or equip himself to take its helm and steer it, but he weighs upon it by the very mass of his assembled presence: being-there extends from Milan to Dublin. If the vanquished acquires a dignity that the victor loses, this is a sign that our world is becoming noble.

The Great Wall of China, it is said, can be seen from the Moon; through growth and dense clustering, we have thus gone beyond a critical size, so that Pascal's points, stuck together, have wound up forming varieties of clusters: surfaces, volumes, masses. Now we are starting to understand the role of great stockpiles in the regulation and evolution of the globe, the specific and combined functions of the seas, the atmosphere, giant deserts, and glaciers. From now on there will be lakes of humanity, physical actors in the physical system of the Earth. Man is a stockpile, the strongest and most connected of nature. He is a being-everywhere. And bound.

According to philosophers of old, men formed a great animal by assembling through a social contract. In the passage from individuals to groups, we rose in size but fell from thought to brute life, brainless or mechanical, so true is it that in saying "we," publicity, meaning the essence of the public, has never really known what it was saying or thinking; such groups may be superior, then, in critical mass, but inferior in the chain of being.

Grazing on green grass or harvested oats, dispersed among the fields and pastures, from time to time looking for someone to devour, this herd of Leviathans, almost as light as being-there, could be neglected in the final balance of the planet's physical system, although it did matter a little in the equilibrium and evolution of the living species to which it belonged: ogres among other monsters.

By growing beyond Leviathan, past a critical mass, the collective moves up from monster to sea, while falling from the living to the inanimate, whether natural or constructed. Yes, the megalopolises are becoming physical variables: they neither think nor graze, they weigh.

Thus the prince, formerly a shepherd of beasts, will have to turn to the physical sciences and become a helmsman or cybernetician.

The relations of man and the world reach their culmination, are transformed, and are even reversed.

Physically nil, a thinking animal lost among species better adapted than his, the individual or being-there has as much effect on the world as a butterfly in the Australian desert, whose beating wings will have repercussions over the meadows of green Erin, perhaps tomorrow or in two centuries, in the form of a storm or a gentle breeze, as luck will have it. The *ego* of the *cogito* has the same power and the same remote causality or impact as this trembling lepidopteran wing; thought is equivalent to the stridulating elytrons of a chirring cricket. Let us say that it is equipotent to this scale of events: not more, but not less. While it can happen, improbably, that it unleashes from a distance the power of a cyclone, its effect remains most often—even always, except for the rarest of exceptions—nil. Thought is nothing or tremendously powerful, it all depends.

To be sure, the local causal chain becomes more effective when thought is limited to plans for raising a stone wall or taming a plow ox. But nothing in such plans concerns global nature, which alone is decisive today.

The entire history of science consists of controlling and mastering this chain, of making consistent the highly improbable linkage of butterfly thought to hurricane effect. And the passage from this soft cause to these hard consequences precisely defines contemporary globalization.

Still physically nil, the old-style group, a living Leviathan, had merely biological efficacy and merely brute thought. By means of the great animal, we have so fully won the struggle for life against other species of flora and fauna that, having reached a threshold, we fear that victory, suddenly, will be overturned into defeat.

Finally we have reached such sizes that we exist physically. The thinking individual, having become a beast collectively, is now joined to others in multiple ways and turns to stone. Upon this rock is built the new world. The hard, hot architecture of megalopolises is equal to many a desert, to groups of springs, wells, lakes—far greater streams than the river of Achilles, shifting sands so much larger than Goya's quicksand—or to an ocean, or a rigid and mobile tectonic plate. At last we exist on a natural scale. Mind has grown into a beast and the beast is growing into a plate.

Henceforth we take up the entire chain of beings, spiritual, living, and inanimate: I think as an individual; we used to live as collective animals; our collectivities are becoming as powerful as seas and share the same destiny. We have invaded not only the space of the world, but, if I dare say so, ontology. First in thought or communication, the best informed of organized beings, the most active of material totalities. Being-everywhere spreads not only in extension, but from one kingdom of being to another.

My cogitating butterfly-wing causality is coupled with our vital effects on species, and now it reaches the stage of purely physical action. In any case, I used to be and obviously still am a local player in the hard and soft sciences; from now on I am also an agent with a remote chance of having effects on a global scale measurable by the physical sciences; but collectively we have powerful and weighty effects on the entire world of all the natural sciences. Fragility has just changed sides.

This is who is found on the fourth corner of the square or at the very end of the gyroscopic axis: being-in-the-world transformed into a being as powerful as the world.

And this equipotency is making the fight's outcome doubtful.

The new counterpart of these new plates of humanity is global nature, Planet Earth in its totality, the seat of reciprocal and crossed interrelations among its local elements and its giant components—oceans, deserts, atmospheres, or stocks of ice. The human plates themselves are the seats of reciprocal and crossed interrelations among individuals and subgroups, their tools, their world-objects, and their knowledge, assemblages that little by little are losing their relations with place, locality, neighborhood, proximity. Being-there is getting rarer.

This is the state, the balanced account, of our relations with the world, at the beginning of a time when the old social contract ought to be joined by a natural contract. In a situation of objective violence, there is no way out but to sign it.

At the very least, war; ideally, peace.

Knowing

Solitude slides so quickly toward inventive delirium and error that the site of knowledge production is never a relation between an individual and his object, but rather one between a growing body of researchers checking on one another and a carved out specialty, defined and accepted by them.

With the origin of science, the former imaginary subject of knowledge, taking refuge in his stove-heated room to conjure up the Devil and the Good Lord, or bent under his transcendental conditions, gives way to a group, united or dispersed in space and time, dominated and ruled by an agreement. This agreement has been said to be consensual, or else, on the contrary, to be endlessly traversed by polemics and debates. Both are true, depending on the scene of knowledge or the historical moment, and those who fight contract to agree, even more here than in the case discussed previously.

This war or this peace, in sum, is based on a tacit contract resembling the old social contract, and it brings together scientists, like the refined debaters, the soldiers, or the economic rivals of a moment ago. Before this tacit contract there was no more science than there was society before the social contract. At the most distant Greek origins of the most rigorous thought, the first scientists, whether assembled or dispersed, debated even more than they proved, jurists as much as geometers.

The subject of knowledge, thus defined as the bond uniting the participants in the scientific enterprise, is much less a matter of a common oral or written language, fluctuating and varied, than has commonly been believed. It amounts, rather, to a tacit and stable contract behind or under this language, a contract whose legal subject is the subject of science: virtual, current, formal, operational.

Let's just list the successive incarnations of this subject: beginning in infancy, the individual enters into relation with the community, which is already bound by this contract; well before starting to examine the objects of his specialization, he presents himself before accredited examining boards, which decide whether or not to receive him among the learned; after having learnedly worked, he presents himself once again before other authorities, who decide whether or not to receive his work into their canonized language. There can be no knower without the first judgment, no knowledge without the second. Thus experienced by the former individual subject, me or you, an obedient receiver or transmitter and a possible inventive producer of knowledge, the process of knowing runs from trials to cases to causes, from judgments to choices, and so never leaves the juridical arena. The sciences proceed by contracts. Scientific certainty and truth depend, in fact, as much on such judgments as such judgments do on them.

The history of the sciences often merges with that of the pronouncements of courts—or of authorities, scientific and otherwise: this will become abundantly clear. The knowledge recognized as scientific ensues from this "epistemodicy"; I mean by this new word all relations of science and law, reason and judgment.

The tribunals of knowledge know causes, which are often conflictual, before knowing things, which are often peaceful, even if scientists know things before fighting about causes. In science, law anticipates fact as subjects precede the object; but fact anticipates law as object precedes subject.

Thus the legal contract that brings together scientists involves things; it discovers them, analyzes them, and constitutes them as scientific objects. Once again a worldly world regulated by contract enters into relation with the worldwide world regulated by laws of a different kind. We don't know how to describe the relations between these laws and the juridical laws of courts, which take cognizance of our causes or cases.

In other words, scientific knowledge results from the passage that changes a cause into a thing and a thing into a cause, that makes a fact become a law, de facto become de jure, and vice versa. The reciprocal transformation of cause into thing and of law into fact explains the double situation of scientific knowledge, which is, on the one hand, arbitrary convention, as is all speculative theory, and, on the other hand, the faithful and exact objectivity that underlies every application.

Consequently, the relation of law to fact, of contract to world, which we noted in dialogue, rivalry, and conflicts, renews itself unchanged in scientific knowledge: by definition and in its real functioning, science is an ongoing relation between the contract uniting scientists and the world of things. And this relation between convention and fact, unique in human history, so miraculous that since Kant and Einstein we have not ceased to marvel at it, has not been given a juridical name. It is as if the verdicts of humans coincide with those of objects. That never happens, except in miracles and sciences.

We're talking about a law, thus arbitrary convention. But it concerns facts, established and checked—those of nature. So since its establishment, science has played the role of natural law. This time-honored expression conceals a profound contradiction, that of the arbitrary and the necessary. Science conceals the same contradiction, in exactly the same places. Physics is natural law: it has played this role since its dawning. The cardinals who defended natural law were beaten at their own game by Galileo, who held to physics.

Who then can be surprised that the question of natural law today depends closely on science, which also describes the place of groups in the world? For, what is more, the scientific collectivity, a minuscule subset of the vast human plate, finds itself facing other collectivities with which it maintains classical relations, consensual or aggressive, to be determined by ordinary contracts.

Consequently, the basic combat situation reappears in knowledge. There, just as we noted previously, a collectivity united by an agreement finds itself facing the world in a relation, neither dominated nor managed, of unconscious violence: mastery and possession.

The origin of science resembles the origin of human societies as if they were sisters: the pact of knowing, a type of social contract, cooperatively controls the expressions of knowledge. But this pact does not make peace with the world, even though it is closer to it.

Why should we be surprised today to hear contradictory arguments about the beneficial or damaging effects of knowledge or reason, which has itself been passing judgments for more than two millennia? More than three hundred years ago, a much-vaunted *Theodicy* decided on the cause of sufferings and evil and concluded that the Creator was tragically responsible. We do not know before what court or in what forms to argue a similar case, a case where it is once more a matter of good and evil, but where the rational producer and far-sighted person in charge has long since rejoined the human collectivity. *Epistemodicy,* that is an exact and possible title of this book, though it is too ugly to be adopted.

Science brings together fact and law: whence its now decisive place. Scientific groups, in a position to control or do violence to the worldwide world, are preparing to take the helm of the worldly world.

Beauty

The very being of beauty, nothing is as beautiful as the world; nothing beautiful comes forth without this gracious giver of all splendor. Amid the atrocities of the Trojan War, blind Homer sings of rosy-fingered dawn; from the pride of bulls comes the strength of Goya, whose paintings bemoan similar and more recent horrors. To anyone who detaches himself from battles because even average wisdom makes them seem vain, if not inhuman, or who does not want to pay for his worst desires with infamy, the worldwide world today offers the painful face of mutilated beauty. Will the strange and timid radiance of dawn be harmed by our brutality?

Out of the equivalence, the identity, the fusion of the worldwide world and the worldly world arises beauty. Thus it surpasses the real in the direction of the human and the human in the direction of the real, and in both cases sublimates both. Epistemology and esthetics, the latter in both its meanings, held forth about the harmony of the rational and the real without being able to explain this miracle, which, I repeat, astounded Kant and Einstein, others too, and left them speechless.

From an old word of the sacred tongue, which signified stain and profanation, insult, violation, and dishonor, we call the breakdown of this equipollence *pollution.* How have divine landscapes, the saintly mountain and the sea with the innumerable smiles of the gods, how have they been transformed into sewage farms or horrifying dumping grounds for corpses? By scattering material and sensory garbage, we are covering or erasing the world's beauty and reducing the luxurious proliferation of its multiplicities to the desert and solar uniformity of our laws alone.

More terrifying than the purely speculative probability of a flood, such a wave of poisons poses, though in inverse form, the same problem of history, law, philosophy, even of metaphysics, that beauty's enigma posed not long ago. In bygone days, the equivalence or meeting of two worlds, song of harmony and elation, marked the optimism and happiness of our ancestors—amid the horror of combats or debates, no one could deprive them of the world—just as now the rupture of these two worlds awakens our anxiety.

If our rational could wed the real, the real our rational, our reasoned undertakings would leave no residue; so if garbage proliferates in the gap between them, it's because that gap produces pollution, which fills in the distance between the rational and the real. Since the filth is growing, the breach between the two worlds must be getting worse. Ugliness ensues from discord and vice versa. Do we still have to prove that our reason is doing violence to the world? Does our reason no longer feel the vital need for beauty?

Beauty demands peace; peace depends on a new contract.

Peace

The only strong or concrete reason that peoples and states have found to join forces and institute a lasting truce among themselves is the formal idea of perpetual peace, an idea that has always been abstract and inconsequential because nations have been able to consider themselves, as a group, alone in the world. Nothing and nobody and no collectivity was above them, and thus no reason.

Since the death of God, all we have left is war.

But now that the world itself is entering into a natural contract with the assembled peoples, however conflictual their assembly may be, it gives the reason for peace, as well as the sought after transcendence.

We must decide on peace among ourselves to protect the world, and peace with the world to protect ourselves.

Natural Contract

Time and Weather

By chance or wisdom, the French language uses a single word, *temps,* for the time that passes and for the weather outside, a product of climate and of what our ancestors called meteors.

Today our expertise and our worries turn toward the weather, because our industrious know-how is acting, perhaps catastrophically, on global nature, which those same ancestors thought didn't depend on us. From now on, not only does it doubtless depend on us, but, in return, our lives depend on this mobile atmospheric system, which is inconstant but fairly stable, deterministic and stochastic, moving quasi-periodically with rhythms and response times that vary colossally.

How are we adding to its variation? What serious disequilibria will occur, what global change must be expected in the whole climate from our growing industrial activities and technological prowess, which pour thousands of tons of carbon monoxide and other toxic wastes into the atmosphere? As of now we don't know how to estimate general transformations on such a scale of size and complexity. Above all, we surely don't know how to think about the relations between time and weather, *temps* and *temps:* a single French word for two seemingly disparate realities. For do we know a richer and more complete model of global change, of equilibria and their attractors, than that of climate and the atmosphere? We are trapped in a vicious circle.

In other words: what risks are we running? Above all: beyond what threshold and at what date or temporal limit will a major risk appear? Since, for the time being, we don't know the answers to these questions, prudence—and politicians—are asking: what to do? when to do it? how and what to decide?

First of all: who will decide?

Peasant and Sailor

In days gone by, two men lived out in the often intemperate weather: the peasant and the sailor. How they spent their time, hour by hour, depended on the state of the sky and on the seasons. We've lost all memory of what we owe these two types of men, from the most rudimentary technologies to the highest subtleties. A certain ancient Greek text divides the earth into two zones: one where a given tool was regarded as a grain shovel and another where passersby recognized the same tool as an oar. In the West, these two populations are gradually disappearing from the face of the earth; agricultural surpluses and high-tonnage vessels are turning the sea and the land into deserts. The greatest event of the twentieth century incontestably remains the disappearance of agricultural activity at the helm of human life in general and of individual cultures.

Now living only indoors, immersed only in passing time and not out in the weather, our contemporaries, packed into cities, use neither shovel nor oar; worse yet, they've never even seen them. Indifferent to the climate, except during their vacations when they rediscover the world in a clumsy, arcadian way, they naively pollute what they don't know, which rarely hurts them and never concerns them.

Dirty species, monkeys and motorists, drop their filth fast, because they don't live in the space they pass through and thus let themselves foul it.

Once again: who decides? Scientists, administrators, journalists. How do they live, and, more important, where? In laboratories, where the sciences reproduce phenomena to define them better; in offices or studios. In short, indoors. The climate never influences our work anymore.

How do we keep ourselves busy? With numerical data, equations, dossiers, legal texts, news bulletins hot off the press or the wire: in short, language. True language for science, normative language for administrators, sensational language for the media. From time to time some expert, a climatologist or an earth scientist, goes off on official business to gather on-site observations, like some reporter or inspector. But the essentials take place indoors and in words, never again outdoors with things. We've even walled up the windows in order to hear one another better or argue more easily. We communicate irrepressibly. We busy ourselves only with our own networks.

Those who share power today have forgotten nature, which could be said to be taking its revenge but which, more to the point, is reminding us of its existence, we who live in time but never right out in the weather. We claim to talk meaningfully about the weather; now we have to make decisions about it.

We have lost the world. We've transformed things into fetishes or commodities, the stakes of our stratagems; and our a-cosmic philosophies, for almost half a century now, have been holding forth only on language or politics, writing or logic.

At the very moment when we are acting physically for the first time on the global Earth, and when it in turn is doubtless reacting on global humanity, we are tragically neglecting it.

Long Term and Short Term

But even if we are reduced to living only in time, not in weather—in which time, once again, are we living? The universal answer today: in the very short term. To safeguard the earth or respect the weather, the wind and rain, we would have to think toward the long term, and because we don't live out in the weather, we've unlearned how to think in accordance with its rhythms and its scope. Concerned with maintaining his position, the politician makes plans that rarely go beyond the next election; the administrator reigns over the fiscal or budgetary year, and news goes out on a daily and weekly basis. As for contemporary science, it's born in journal articles that almost never go back more than ten years; even if work on the paleoclimate recapitulates tens of millennia, it goes back less than three decades itself.

It is all as if the three contemporary powers—and by powers I mean those institutions that don't run up against counterpowers anywhere—had eradicated long-term memory, the thousand-year-old traditions, the experience accumulated by cultures that have just died or that these powers are killing.

Here we are faced with a problem caused by a civilization that has now been in place for more than a century and that was itself engendered by the long-term cultures that preceded it; it's damaging a physical system millions of years old, a system that fluctuates and yet remains relatively stable through rapid, random, and multicentury variations. Before us is an anguishing question, whose principal component is time, especially a long-term time that is all the longer when the system is considered globally. Mixing the waters of the oceans requires a cycle estimated at five thousand years.

We are proposing only short-term answers or solutions, because we live with immediate reckonings, upon which most of our power depends. Continuity belongs to administrators, the day-by-day to the media, and to science belong the only plans for the future we have left. The three powers have control over time, so now they can rule or decide on the weather.

How can one not be surprised, by the way, by the situation in news reporting, where the reduction of time to the all-important passing instant is paralleled by the obligatory reduction of news items to passing catastrophes, assumed to be the only things of interest? It's as if the very short term were linked to destruction: must we understand, conversely, that building requires the long term? The same thing in science: what secret relations does refined specialization have with analysis, which destroys the object, already carved up by specialization?

It so happens that we must decide about the greatest object of scientific knowledge and practice, the Planet Earth, this new nature.

To be sure, we can slow down the processes already under way, legislate reductions in fossil-fuel consumption, massively replant the devastated forests . . . all fine initiatives, but together they amount to the image of a ship sailing at twenty-five knots toward a rocky bar on which it will inevitably be smashed to pieces, and on whose bridge the officer of the watch advises the engine room to reduce speed by a tenth without changing direction.

To become effective, the solution to a long-term, far-reaching problem must at least match the problem in scope. Those who used to live out in the weather's rain and wind, whose habitual acts brought forth long-lasting cultures out of local experiences—peasants and sailors—have had no say for a long time now, if they ever had it. It is we who still have a say: administrators, journalists, and scientists, all men of the short term and of highly focused specialization. We're partly responsible for the global change in the weather, because we invented or distributed the means and the tools of powerful, effective, beneficent, and harmful intervention; we're inept at finding reasonable solutions because we're immersed in the brief time of our powers and imprisoned in our narrow domains.

If there is a material, technological, and industrial pollution, which exposes weather to conceivable risks, then there is also a second pollution, invisible, which puts time in danger, a cultural pollution that we have inflicted on long-term thoughts, those guardians of the Earth, of humanity, and of things themselves. If we don't struggle against the second, we will lose the fight against the first. Who today can doubt the cultural nature of what Marxists used to call the base?

How are we to succeed in a long-term enterprise with short-term means? We must pay the price of such a plan by a harrowing revision of today's culture, which rests on the three powers that dominate our short-term obsessions. Have we lost the memory of the antediluvian age when a patriarch, from whom we're doubtless descended, had to construct an ark, a small-scale model of the totality of space and time, to prepare for an overflow of the sea caused by some thaw in the ice caps?

In memory of those who have fallen silent forever, let us then give long-term men their say: a philosopher can still learn from Aristotle, a jurist does not find Roman law too old. Let's listen to them a moment, before painting the portrait of the new political leader.

The Philosopher of Science

asks: but who, then, is inflicting on the world, which is henceforth a common objective enemy, this harm that we hope is still reversible, this oil spilled at sea, this carbon monoxide spread in the air by the millions of tons, these acidic and toxic chemicals that come back down with the rain . . . whence comes this filth that is choking our little children with asthma and covering our skin with blotches? Who, beyond private and public persons? What, beyond enormous metropolises, considered either as aggregations of individuals or as networks of relations? Our tools, our arms, our efficacy, in the end our reason, about which we're so legitimately vain: our mastery and our possessions.

Mastery and possession: these are the master words launched by Descartes at the dawn of the scientific and technological age, when our Western reason went off to conquer the universe. We dominate and appropriate it: such is the shared philosophy underlying industrial enterprise as well as so-called disinterested science, which are indistinguishable in this respect. Cartesian mastery brings science's objective violence into line, making it a well-controlled strategy. Our fundamental relationship with objects comes down to war and property.

War, Once Again

The sum total of harm inflicted on the world so far equals the ravages a world war would have left behind. Our peacetime economic relations, working slowly and continuously, produce the same results as would a short global conflict, as if war no longer belonged to soldiers alone now that it is prepared and waged with devices as scientific as those used by civilians in research and industry. Through a kind of threshold effect, the growth of our means makes all ends equal.

We so-called developed nations are no longer fighting among ourselves; together we are all turning against the world. Literally a world war, and doubly so, since the whole world, meaning all men, imposes losses on the world, meaning things. We shall thus seek to conclude a peace treaty.

Dominate, but also *possess:* the other fundamental relationship we have with the things of the world comes down to property rights. Descartes's master word amounts to the application of individual or collective property rights to scientific knowledge and technological intervention.

The Clean and the Dirty

I've often remarked that, just as certain animals piss on their territory so that it stays theirs, many men mark and dirty the things they own by shitting on them, in order to keep them, or shit on other things to make them their own. This stercoraceous or excremental origin of property rights seems to me a cultural source of what we call pollution, which, far from being an accidental result of involuntary acts, reveals deep intentions and a primary motivation.

Let's have lunch together: when the salad bowl is passed, all one of us has to do is spit in it and it's all his, since no one else will want any more of it. He will have polluted that domain and we will consider dirty that which, being clean only to him, he now owns. No one else ventures again into the places devastated by whoever occupies them in this way. Thus the sullied world reveals the mark of humanity, the mark of its dominators, the foul stamp of their hold and their appropriation.

A living species, ours, is succeeding in excluding all the others from its niche, which is now global: how could other species eat or live in that which we cover with filth? If the soiled world is in danger, it's the result of our exclusive appropriation of things.

So forget the word *environment,* commonly used in this context. It assumes that we humans are at the center of a system of nature. This idea recalls a bygone era, when the Earth (how can one imagine that it used to represent us?), placed in the center of the world, reflected our narcissism, the humanism that makes of us the exact midpoint or excellent culmination of all things. No. The Earth existed without our unimaginable ancestors, could well exist today without us, will exist tomorrow or later still, without any of our possible descendants, whereas we cannot exist without it. Thus we must indeed place things in the center and us at the periphery, or better still, things all around and us within them like parasites.

How did the change of perspective happen? By the power and for the glory of men.

Reversal

Through our mastery, we have become so much and so little masters of the Earth that it once again threatens to master us in turn.

Through it, with it, and in it, we share one temporal destiny. Even more than we possess it, it will possess us, just as it did in the past, when old necessity, which submitted us to natural constraints, was still around, but it will possess us differently than back then. Then locally, now globally.

Why must we now seek to master our mastery? Because, unregulated, exceeding its purpose, counterproductive, pure mastery is turning back on itself. Thus former parasites have to become symbionts; the excesses they committed against their hosts put the parasites in mortal danger, for dead hosts can no longer feed or house them. When the epidemic ends, even the microbes disappear, for lack of carriers for their proliferation.

Not only is the new nature, as such, global, but it reacts globally to our local actions.

We must then change direction and abandon the heading imposed by Descartes's philosophy. Because of these crossed interactions, mastery only lasts for the short term before turning into servitude; property, similarly, has a rapid ascendancy or else ends with destruction.

This is history's bifurcation: either death or symbiosis.

Yet this philosophical conclusion, once known and practiced by agrarian and maritime cultures, though on a local scale and in narrow temporal limits, would remain a dead letter if it were not inscribed in law.

The Jurist. Three Laws without a World

THE SOCIAL CONTRACT. Philosophers of modern natural law sometimes trace our origin to a social contract that we are said to have signed among ourselves, at least virtually, in order to enter into the collectivity that made us the men we are. Strangely silent about the world, this contract, they say, made us leave the state of nature to form society. From the time the pact was signed, it is as if the group that had signed it, casting off from the world, were no longer rooted in anything but its own history.

This sounds like the local and historical description of the rural exodus toward the cities. It clearly means that from that point on we forgot the aforementioned nature, which is now distant, mute, inanimate, isolated, infinitely far from cities or groups, from our texts, and from publicity. By "publicity" I mean the essence of the public, which henceforth constitutes the essence of men.

NATURAL LAW. The same philosophers call natural law a collection of rules said to exist outside of any formulation; being universal, this law would follow from human nature. The source of man-made laws, natural law follows from reason inasmuch as reason governs all men.

Nature is reduced to human nature, which is reduced to either history or reason. The world has disappeared. Modern natural law is distinguished from classical natural law by this nullification. Self-important men are left with their history and their reason. Curiously, reason acquires in the legal sphere a status quite similar to the one it had acquired in the sciences: the laws are all on its side because it founds law.

The Declaration of the Rights of Man

In France we celebrated the bicentennial of the Revolution, and at the same time, that of the Declaration of the Rights of Man. Those rights, according to the text, issue explicitly from natural law.

Like the social contract, the Declaration ignores the world and remains silent about it. We no longer know the world because we have conquered it. Who respects victims? The aforementioned Declaration was pronounced in the name of human nature and in favor of the downtrodden and the wretched—those who, excluded, lived outdoors, outside, submerged with all hands in the winds and rain, those for whom the time of life passing was bound to the weather. In the name of those who enjoyed no rights, losers of all imaginable wars who owned nothing.

Monopolized by science and by all the technologies associated with property rights, human reason conquered external nature in a combat that has lasted since prehistory, but that sped up relentlessly with the industrial revolution, more or less at the same time as the revolution whose bicentennial we celebrated, the one technological, the other political. Once again, we must rule in the case of the losers, by drafting the rights of beings who have none.

We conceive law as based on a legal subject, whose definition has progressively broadened. In the past not just anyone could attain this status: the Declaration of the Rights of Man and of the Citizen gave the possibility to every man to attain the status of subject of the law. The social contract was thereby completed, but closed upon itself, leaving the world on the sidelines, an enormous collection of things reduced to the status of passive objects to be appropriated. Human reason was of age, external nature a minor. The subject of knowledge and action enjoys all rights and its objects none. They have not yet attained any legal dignity. Which is why, since that time, science has all the laws on its side.

Thus we necessarily doom the things of the world to destruction. Mastered and possessed, from the epistemological point of view; minors in the pronouncement of the law. Yet they receive us as hosts, and without them we would die tomorrow. Exclusively social, our contract is becoming poisonous for the perpetuation of the species, its global and objective immortality.

What is nature? First, all the conditions of human nature itself, its global constraints of rebirth or extinction, the hostelry that gives us lodging, heat, and food. But nature also takes them away from us as soon as we abuse them. It influences human nature, which, in turn, influences nature. Nature behaves as a subject.

Use and Abuse: The Parasite

In its very life and by its practices, the parasite routinely confuses use and abuse; it accords itself rights, which it exercises by harming its host, sometimes without any advantage for itself. The parasite would destroy the host without realizing it. Neither use nor exchange has value in its eyes, for it appropriates things—one could say that it steals them—prior to use or exchange: it haunts and devours them. The parasite is always abusive.

Conversely, law in general can surely be defined as the minimal and collective limitation of parasitic action. For parasitism, in fact, follows the simple arrow of a flow moving in one direction but not the other, in the exclusive interest of the parasite, which takes everything and gives back nothing along this one-way street. The judicial, on the other hand, invents a double, two-way arrow that seeks to bring flows into balance through exchange or contract;

at least in principle, it denounces one-sided contracts, gifts without countergifts, and ultimately all abuses. The just scale of the law contravenes the parasite from its foundation: this scale sets the equilibrium of a balance sheet against any abusive disequilibrium.

What is justice if not this double arrow, this very scale, or the unbroken effort to institute it within relations of force?

We must therefore carry out a harrowing revision of modern natural law, which presupposes the unformulated proposition that only man, individually or in groups, can become a legal subject. This is where parasitism reappears. The Declaration of the Rights of Man had the merit of saying "every man" and the weakness of thinking "only men," or men alone. We have not yet set up a scale in which the world is taken into account in the final balance sheet.

Objects themselves are legal subjects and no longer mere material for appropriation, even collective appropriation. Law tries to limit abusive parasitism among men but does not speak of this same action on things. If objects themselves become legal subjects, then all scales will tend toward an equilibrium.

Equilibria

There are one or several natural equilibria, described by physical mechanics, thermodynamics, the physiology of organisms, ecology, or systems theory; cultures have even invented one or more human and social equilibria, which are decided on, organized, and maintained by religion, laws, or politics. But something is missing: we are not conceiving, constructing, or putting into operation a new global equilibrium between the two sets of equilibria.

Social systems, which are self-compensating and self-enclosed, press down with their new weight, that of their relations, objectworlds, and activities, on self-compensated natural systems, just as in the past natural systems put social systems at risk, in the age when necessity triumphed over reason's means.

Blind and mute, natural fatality neglected, back then, to sign an explicit contract with our ancestors, whom it crushed: now we are sufficiently avenged for this archaic abuse by a reciprocal modern abuse. It remains to us to imagine a new, delicate balance between these two sets of balances. As far as I know, this just weighing is at the origin of the verb "to think" (*penser*), as well as of the verb "to compensate." Today this is what we name thought. This is the most general legal order for the most global systems.

The Natural Contract

Henceforth, men come back into the world, the worldly into the worldwide, the collectivity into the physical. It's a bit like the era of classical natural law, but with big differences, all of which have to do with the recent passage from the local to the global and with our renewed relationship to the world, which was long ago our master and of late our slave, always and in all cases our host, and now our symbiont.

Back to nature, then! That means we must add to the exclusively social contract a natural contract of symbiosis and reciprocity in which our relationship to things would set aside mastery and possession in favor of admiring attention, reciprocity, contemplation, and respect; where knowledge would no longer imply property, nor action mastery, nor would property and mastery imply their excremental results and origins. An armistice contract in the objective war, a contract of symbiosis, for a symbiont recognizes the host's rights, whereas a parasite—which is what we are now—condemns to death the one he pillages and inhabits, not realizing that in the long run he's condemning himself to death too.

The parasite takes all and gives nothing; the host gives all and takes nothing. Rights of mastery and property come down to parasitism. Conversely, rights of symbiosis are defined by reciprocity: however much nature gives man, man must give that much back to nature, now a legal subject.

What do we give back, for example, to the objects of our science, from which we take knowledge? Whereas the farmer, in bygone days, gave back, in the beauty that resulted from his stewardship, what he owed the earth, from which his labor wrested some fruits. What should we give back to the world? What should be written down on the list of restitutions?

In the last century we pursued the ideal of two revolutions, both egalitarian. The people takes back its political rights, restored after having been stolen; likewise the proletariat regains the enjoyment of the material and social fruits of its labor. These were quests for balance and equity within the exclusively social contract, which was formerly unjust or one-sided, and is always tending to become unbalanced again. So fiercely does our animality strive to restore hierarchy that such a quest can never end. While we're pursuing it, a second quest begins, one that will characterize our future history just as the previous one left its mark on the last century: this second quest involves the same search for balance and justice, but between new partners, the global collectivity and the world as such.

We shall no longer gain knowledge, in the scientific sense, nor will our industries labor and transform the peaceful face and entrails of the world, as we once did: collective death is seeing to this global contractual change.

It could be said that the reign of modern natural law began at the same time as the scientific, technological, and industrial revolutions, with the mastery and possession of the world. We imagined that we'd be able to live and think among ourselves, while things around us obediently slumbered, crushed by our hold on them: human history could take pleasure in itself in an a-cosmism of inanimate matter and of other living things. History can be made of everything and everything comes down to history.

Slaves never sleep for long. This period is coming to an end, now that awareness of things is violently calling us back. Irresponsibility only lasts through childhood.

What language do the things of the world speak, that we might come to an understanding with them, contractually? But, after all, the old social contract, too, was unspoken and unwritten: no one has ever read the original, or even a copy. To be sure, we don't know the world's language, or rather we know only the various animistic, religious, or mathematical versions of it. When physics was invented, philosophers went around saying that nature was hidden under the code of algebra's numbers and letters: that word *code* came from law.

In fact, the Earth speaks to us in terms of forces, bonds, and interactions, and that's enough to make a contract. Each of the partners in symbiosis thus owes, by rights, life to the other, on pain of death.

All of that would remain a dead letter if we didn't invent a new type of political leader.

The Political

When he talks politics, Plato sometimes cites the example of the ship, of the crew's submission to the helmsman, an expert governor, but he never mentions what is exceptional about this model, doubtless because he's unaware of it.

Between ordinary life on land and the heaven or hell at sea lies the possibility of withdrawal: on board, social existence never ceases, and no one can retire to his private tent, as the infantry warrior Achilles did long ago. On a boat, there's no refuge on which to pitch a tent, for the collectivity is enclosed by the strict definition of the guardrails: outside the barrier is death by drowning. This total social state, which delighted the philosopher for reasons we would judge base, holds seagoers to the law of politeness, where "polite" means *politic* or *political.* There is localness, being-there, when there is leftover space.

Since remotest antiquity, sailors (and doubtless they alone) have been familiar with the proximity and connection between subjective wars and objective violence, because they know that, if they come to fight among themselves, they will condemn their craft to shipwreck before they can defeat their internal adversary. They get the social contract directly from nature.

Unable to have any private life, they live in ceaseless danger of anger. A single unwritten law thus reigns on board, the divine courtesy that defines the sailor, a nonaggression pact among seagoers, who are at the mercy of their fragility. The ocean threatens them continuously with its inanimate but fearsome strength, seeing to it that they keep the peace.

The social pact of courtesy on the seas is altogether different from the contract by which other human groups organize themselves and even originate. The seagoing pact is in fact equivalent to what I'm calling a natural contract. Why? Because here the collectivity, if sundered, immediately exposes itself to the destruction of its fragile niche, with no possible recourse or retreat. Its habitat has no supplement, no refuge like Achilles' tent, that small private landlubbers' fort to which the light infantryman can retreat when angry with fellow footsoldiers. Because it has no leftover space to which to withdraw, the ship provides a model of globality: being-there, which is local, belongs on land.

From the beginning of our culture, the *Iliad* is opposed to the *Odyssey* as conduct on land, which takes only people into account, is to the ways of the sea, which deal with the world. Thus the soldiers of the first poem, an historical epic, become companions in the second, literally a geographical text and map, where the known Earth itself writes, and where we can already see a natural contract, concluded silently and out of fear or respect, between the rumbling ire of the great social beast and the noise, sound, and fury of the sea. An agreement between deaf Ulysses and the clamoring Sirens, a pact between the prow and the waves, the peace of men confronting the winds. But what language do the things of the world speak? The voice of the elements comes through the throats of those strange women who sing in the enchanted straits.

We have made politics and economics into their own disciplines so as to define power: how are we to think of fragility? By the absence of a supplement. In contrast, strength has reserves at its disposal; it defends itself elsewhere, attacks along other lines, falls back on prepared positions, like Achilles in his tent, and has provisions to eat. A full and rigid totality, however, can break from rigor or hardness, like a prow lying to against surging waves. This explains why fuzzy sets, equipped with leftover spaces and refuges, can be so resistant.

There's nothing weaker than a global system that becomes a single unit. A single law corresponds to sudden death. The more plural an individual becomes, the better he lives: the same is true for societies, or for being in general.

Here, then, is the form of contemporary society, which can be called doubly worldwide: occupying all the Earth, solid as a block through its tightly woven interrelations, it has nothing left in reserve, no external place of withdrawal or recourse on which to pitch its tent. Society knows, moreover, how to construct and use technologies whose spatial, temporal, and energetic dimensions are on the scale of worldwide phenomena. Our collective power is therefore reaching the limits of our global habitat. We're beginning to resemble the Earth.

Thus our de facto unified group borders the world, to which it is equipotent, just as the solid and mobile deck all but touches the surface of the waves, separated only by the stanchions of the guardrail. Everyone sails upon the world like the ark upon the waters, without any reserve outside these two sets, that of men and that of things. So here we are, underway! For the first time in history, Plato and Pascal, who never went to sea, are both right at the same time, for here we are, constrained to obey shipboard laws, to pass from the social contract to the natural contract. The social contract long protected mobile social subsets in a broad and free environment, equipped with reserves that could absorb any damage, but a unified, compact group that has reached the strict limits of objective forces requires a natural contract.

There our global-range arms and technologies affect the totality of the world, whose wounds thus inflicted affect, in return, the totality of human beings. The objects of politics are henceforth these three connected totalities.

Of Governing

The helmsman governs. Following his intended route and according to the direction and force of the sea-swell, he angles the blade of the governail, or rudder. His will acts on the vessel, which acts on the obstacle, which acts on his will, in a series of circular interactions. First and then last, first a cause and then a consequence, before once again becoming a cause, the project of following a route adapts in real time to conditions that unceasingly modify it, but through which it remains stubbornly invariant. The helmsman's project decides on a subtle and fine tilt of the rudder, a tilt selected within the directional movement of objective forces, so that in the end the route can be traced through the set of constraints.

Cybernetics was the name given to the literally symbiotic art of steering or governing by loops, loops engendered by these angles and that engender, in turn, other directional angles. This technique was once specific to helmsmen's work, but it has recently passed into other technologies just as intelligent as this command of seaworthy vessels; it has moved from this level of sophistication to the grasping of even more general systems, which could neither subsist nor change globally without such cycles. But this whole arsenal of methods remained only a metaphor when it came to the art of governing men politically. What does the helmsman with his governail have to teach those who govern?

Their difference is now vanishing. Today, what everyone does gives rise to harm inflicted on the world, and this damage, through an immediate or foreseeably deferred feedback loop, becomes the givens of everyone's work. I am deliberately playing on a single word of exchange, *give:* we receive gifts from the world and we inflict upon it damage that it returns to us in the form of new givens.

Cybernetics is back. For the first time in history, the human or worldly world is united in facing the worldwide world, without play, remainder, or recourse for the whole of the system, just as on a ship. The governor and the helmsman with his governail become identified in a single art of governing.

The helmsman acts in real time, here and now, on a local circumstance from which he counts on obtaining a global result; it is the same for the governor, the technician, and the scientist. When scientists, gathering their local models into a totality mimicking the Earth, plan some intervention, they speak of steering committees and pilot projects.

Immersed in the exclusively social contract, the politician has been countersigning it up until today, rewriting it, and having it observed: he is solely an expert in public relations and social sciences. Eloquent, even an orator, conceivably a man of culture, he knows minds and hearts, and group dynamics; there's quite a bit of administrator in him, and of media personality—a must. Essentially he's a jurist, a product and producer of law. No point in trying to make himself a physicist.

None of his speeches spoke of the world: instead they endlessly discussed men. Once again, publicity, as the rules for forming such a word demand, is defined as the essence of the public: thus, more than anyone else, the politician indulges in no speech or gesture without saturating them in publicity. What's more, recent history and tradition taught him that natural law expresses only human nature. Closed up in the social collectivity, he could be splendidly ignorant of the things of the world.

All that has just changed. The word *politics* must now be considered inaccurate, because it refers only to the *polis,* the city-state, the spaces of publicity, the administrative organization of groups. Yet those who live in cities, once known as bourgeois, know nothing of the world. From now on, those who govern must go outside of the human sciences, outside the streets and walls of the city, become physicists, emerge from the social contract, invent a new natural contract by giving back to the word *nature* its original meaning of our natal and native conditions, the conditions in which we are born—or ought to be reborn tomorrow.

Conversely, the physicist, in both the most ancient Greek and the most modern sense, is getting closer to the politician.

In a memorable passage describing the art of governing, Plato sketches the king weaving rational weft yarn onto a warp bearing less reasonable passions. Now we are at the dawn of an era in which the new prince will have to cross the woof of law with a warp born of the physical sciences: political art will follow this weaving.

In the past I gave the name Northwest Passage to the place where these two types of knowledge converged, but I did not know, in so doing, that I was defining today's political science, geopolitics in the sense of the real Earth, physiopolitics in the sense that the institutions that groups create will henceforth depend on explicit contracts concluded with the natural world. The natural world will never again be our property, either private or common, but our symbiont.

History, anew

However mythical it may be, the social contract thus marks the beginning of societies. Because of some needs or other, given men decide, on a given day, to live together and thus to join forces; since that time, we no longer know how to do without one another. When, how, why this contract was—or was not—signed, we do not know and will doubtless never know. What does it matter?

Since that legendary time, we have multiplied the number of legal contracts. We can't decide whether these contracts were established on the model of the first one, or whether we imagine the fiction of the originary contract on the model of the standard contracts settled by our laws. Again, what does it matter?

But these laws had and have the genius of delimiting objects, which are attributable by these laws to subjects the laws also define.

We imagine that the social contract joined together, purely and simply, bare individuals, whereas laws, since they involve cases and causes and recognize the existence of things, integrate things into society. Laws thus steady society by using ponderous objects to weigh down fickle subjects and their unstable relationships. No human collectivity exists without things; human relations go through things, our relations to things go through men: this is the slightly more stable space described by laws. Sometimes I imagine that the first legal object was the cord, the bond, *lien* in French, which we read only abstractly in the terms *obligation* and *alliance,* but more concretely in *attachment,* a cord that materializes our relations or changes them into things. If our relations fluctuate, this solidification settles them.

Based on the model of those contracts, a new collectivity joined together, this time on known historical dates, to stabilize objects better still. The contract of scientific truth synthesizes an exclusively intersubjective social contract of constant reciprocal surveillance and agreement in real time about what it is appropriate to say and do, with a truly juridical contract defining certain objects, and specifying competencies, experimental procedures, and the analytical attribution of properties. Then things gradually leave the network of our relations to take on a certain independence; truth demands that we speak of things as if we were not there. A science, from its inception, indissolubly associates the collectivity and the world, the accord and the object of accord.

The contractual act makes these three types of association (social, legal, and scientific) resemble one another, even though the social contract is globally collective, the thousand varieties of law disperse association into a thousand subgroups, and scientific contracts are at once local and global. But the relations of these associations to objects distinguish them. The world, which is totally absent from the social contract, as from the social sciences, is slowly permeating collective decisions: through causes that have become things, then through the causality of things themselves. Only bit by bit does the world make its way into these collectivities. How few philosophies conceive the collectivity as living in the global world!

From now on I mean by natural contract above all the precisely metaphysical recognition, by each collectivity, that it lives and works in the same global world as all the others; not only every political collectivity joined by a social contract but also every other kind of collectivity—military, commercial, religious, industrial, and so on—joined by a legal contract, and also the collectivity of experts joined by the scientific contract. I call the natural contract metaphysical because it goes beyond the ordinary limitations of the various local specialties, physics in particular. It is as global as the social contract and in a way makes the social contract enter the world, and it is as worldwide as the scientific contract and in a way makes the scientific contract enter history.

Virtual and unsigned like the first two contracts, since it seems that the great fundamental contracts remain tacit, the natural contract recognizes and acknowledges an equilibrium between our current power and the forces of the world. Just as the social contract recognizes some equality between its human signatories, just as the various legal contracts seek to balance the interests of the parties, and just as the scientific contract creates an obligation to repay in reason what is received in information, so the natural contract acknowledges above all the new equality between the force of our global interventions and the globality of the world. The thing that stabilizes our relations and that science measures remains local, carved up, limited; law and physics define it. Today it is growing to the dimensions of the Earth.

Finally, the scientific truth contract succeeds brilliantly in showing us the object's point of view, as it were, just as the other contracts showed us, by the bond or ligature of their obligations, so to speak, the point of view of the other partners in the accord. The natural contract leads us to consider the world's point of view in its totality.

Any contract creates a collection of bonds, whose network canonizes relations; today nature is defined by a set of relations whose network unifies the whole Earth. The natural contract connects the second and first networks into one.

The Religious

We are constantly losing our memory of the strange acts that priests used to practice alone in somber and secret nooks, where they would dress the statue of a god, adorn it, ready it, raise it up or take it out, prepare a meal for it and talk to it constantly. They would do this every day and every night, at dawn and at dusk, when the sun and the shade reached their apogee. Were they afraid that a single pause in this continuous, infinite upkeep and conversation would open the door to tremendous consequences?

Amnesiacs that we are, we believe that they adored the god or goddess sculpted in stone or wood. No: they were giving to the thing itself, marble or bronze, the power of speech, by conferring on it the appearance of a human body endowed with a voice. So they must have been celebrating their pact with the world.

Likewise we are forgetting why the Benedictine monks get up before dawn to sing matins and lauds, the minor hours of prime, terce, and sext, or put off their rest until late into the night in order to chant again, at compline. We don't remember these necessary prayers or these perpetual rites. And yet not far from us, Trappists and Carmelites still say the divine office without respite.

They are not following time, but sustaining it. Their shoulders and their voices, from biblical verses to orisons, bear each minute into the next throughout duration, which is fragile and would break without them. And who, conversely, convinces us that there are no gaps in the threads or fabrics of time? Day and night, Penelope never left her loom. In the same way, religion presses, spins, knots, assembles, gathers, binds, connects, lifts up, reads, or sings the elements of time. The term *religion* expresses exactly this trajectory, this review or prolonging whose opposite is called negligence, the negligence that incessantly loses the memory of these strange actions and words.

The learned say that the word religion could have two sources or origins. According to the first, it would come from the Latin verb *religare,* to attach. Does religion bind us together, does it assure the bond of this world to another? According to the second origin, which is more probable, though not certain, and related to the first one, it would mean to assemble, gather, lift up, traverse, or reread.

But they never say what sublime word our language opposes to the religious, in order to deny it: *negligence.* Whoever has no religion should not be called an atheist or unbeliever, but negligent.

The notion of negligence makes it possible to understand our time and our weather.

In the temples of Egypt, Greece, or Palestine, our ancestors, I believe, used to sustain time, as if they were anxious about possible gaps. Here we are today, worried about disasters in the aerial protective fabric that guarantees not time passing, but the weather. They used to connect, assemble, gather, lift up, never ceasing all day long, like monks. And what if it turned out that human history and tradition exist simply because men devoted to the longest term conceivable have never stopped sewing time back together?

Modernity neglects, speaking in absolute terms. It cannot and will not think or act toward the global, whether temporal or spatial.

Through exclusively social contracts, we have abandoned the bond that connects us to the world, the one that binds the time passing and flowing to the weather outside, the bond that relates the social sciences to the sciences of the universe, history to geography, law to nature, politics to physics, the bond that allows our language to communicate with mute, passive, obscure things—things that, because of our excesses, are recovering voice, presence, activity, light. We can no longer neglect this bond.

While we uneasily await a second Flood, can we practice a diligent religion of the world?

Some organisms, it is said, disappeared from the face of the Earth as a result of their enormous size. This still astonishes us, that the biggest things should be the weakest, like the whole Earth, Man as megalopolis or Being-everywhere, even God. Having long enjoyed the death of these grand and fragile entities, philosophy today is taking refuge in small details, which give it a sense of security.

Whose diligent shoulders can henceforth sustain this immense and fissured sky, which we fear, for the second time in a long history, will fall on our heads?

Love

Without love, there are no ties or alliances. Here, finally, are the two double laws.

Love one another, this is our first law. For two thousand years, this is the only one that has been able to help us avoid, at least for a few moments, hell on earth. This contractual obligation is divided into a local law asking us to love our neighbors and a global law requiring us to love at least humanity, if we do not believe in a God.

It is impossible to separate the two precepts, under penalty of hatred. Loving only one's neighbors or one's own kind leads to the team, the sect, to gangsterism and racism; loving men as a group, while exploiting one's neighbors and kin, is the typical hypocrisy of preachy moralists.

This first law remains silent about mountains and lakes, for it speaks to men about men as if there were no world.

Here then is the second law, which asks us to love the world. This contractual obligation is divided into the old local law that attaches us to the ground where our ancestors lie, and a new global law that no legislator, as far as I know, has yet written, which requires of us the universal love of the physical Earth.

It is impossible to separate the two precepts, under penalty of hatred. Loving the whole Earth while laying waste the neighboring landscape, this is the typical hypocrisy of moralists who restrict the law to men and to language, which men alone use and master; loving only one's own land leads to inexpiable wars caused by the passions of belonging.

We used to know how to love our neighbor sometimes, and often the land; we have learned with difficulty to love humanity, which was once so abstract, but which we are starting to encounter more frequently; now we must learn and teach around us the love of the world, or of our Earth, which we can henceforth contemplate as a whole.

Love our two fathers, natural and human, the land and the neighbor; love humanity, our human mother, and our natural mother, the Earth.

It is impossible to separate these two double laws, under penalty of hatred. To defend the land, we have attacked, hated, and killed so many men that some of them came to believe that these killings were the motor of history. Conversely, to defend or attack other men, we have laid waste the landscape without thinking about it, and we were preparing to destroy the whole Earth. So the two contractual obligations, social and natural, have the same solidarity between them as that which binds men to the world and the world to men.

Thus these two laws make up only one, which is inseparable from justice, at once natural and human. Together these laws ask each of us to pass from the local to the global, a difficult and badly marked trail, but one that we must blaze. Never forget the place from which you depart, but leave it behind and join the universal. Love the bond that unites your plot of earth with the Earth, the bond that makes kin and stranger resemble each other.

Peace then between the friends of forms and the sons of the Earth, between those who pronounce the law and those who are attached to the land, peace between the separated brothers, between the idealists of language and the realists of things themselves, and let them love one another.

There is nothing real but love, and no other law.

Science, Law

Origins

EGYPT'S WAY. The first laws on Earth. Given normal weather, the Nile's floods submerged the borders of tillable fields in the alluvial valley fertilized by the great river. At the return of low water, royal officials called *harpedonaptai,* who were surveyors or geometers, measured anew the land mixed with mud and silt to redistribute or attribute its parts. Life got going again. Everyone went home to get back to work.

Floods take the world back to disorder, to primal chaos, to time zero, right back to nature, in the sense of things about to be born, in a nascent state. Correct measurement reorders nature and gives it a new birth into culture, at least in the agricultural sense. If these events give birth to geometry, as Herodotus, who tells this story of emergence, suggests, then geometry has the power of beginning, for the origins of geometry are less important than the geometry of origins.

In another context, it is written in Genesis that God divided the dry land from the primordial waters and gave it form. At the beginning of time, in the same way, the flood's chaotic hubbub is followed by partition: the conditions of definition, measurement, and emergence appear together, departing from chaos. "Departure," which means beginning, also means partition, as I wish to demonstrate.

Deciding on markers and borders indeed appears to be a moment of origin; without such decisions, there is no oasis separate from the desert, no clearing in the forest where peasants set themselves to farming, no sacred or profane space, isolated from each other by priestly gesture, no definition enclosing a domain, and therefore no precise language on which to agree, nor any logic; and finally, no geometry.

But, still more originally, who makes this decision? The term *decision* also expresses cutting, the creation of an edge.

Assigning limits stops disputes among neighbors; the right of property, the right to enclose precisely a plot of ground and attribute it to someone, gives rise to some civil and private law. Moreover, the same delimitation by markers permits the royal cadastre to put each person in his place and to assess the various taxes, which gives rise to public and fiscal law. Without appearing expressly in the *Histories* of Herodotus, laws proliferate in this origin legend, and these laws alone make decisions and divide the fields, whoever the physical person is—the Pharaoh's envoy, the mysterious *harpedonaptes*—who actually restores the fields to their owners. Who decides? The legislator or whoever dictates the law and has it applied.

So this is the person who carries out the foundational act that gives birth to geometry, which will then produce a new agreement among those who do proofs, as if accuracy or justness surpassed even justice in its success. But justice, on this score, came before justness, and identified justness with itself. Before the scientific consensus on the precision of the cut or the need for proof, a legal contract imposes itself and first brings everyone concerned to an accord.

But, once again, since the flood erased the limits and markers of tillable fields, properties disappeared at the same time. Returning to the now chaotic terrain, the *harpedonaptai* redistribute them and thus give new birth to law, which had been erased. Law reappears at the same time as geometry; or rather, both are born along with the notion of limit, edge, and definition, with analytic thought. The definition of precise form implies properties: for geometry, those of the square or the parallelogram; for law, it implies the proprietor. Analytic thought takes root in the same word and the same operation, from which grow two branches, science and law.

The *harpedonaptes* or surveyor draws, holds, and ties the cord;

his mysterious title can be broken down into two words, a noun expressing the bond and a verb denoting his act of attaching it. In the beginning is this cord. The one, for example, that in a temple separates the sacred and the profane. The one evoked by the word *contract*.

The first priest holding this piece of string who, having enclosed a plot of ground, found his neighbors satisfied with the borders of their common enclosure, was the true founder of analytic thought, and thence of law and geometry. That is so because of the permanence of the contract, concluded for a very long time period, because of the exactitude and rigor of the outline, and because of the correspondence between the outline's precision and the contract's stability. The contract is a pact that becomes all the better as its terms are refined, as values are fixed, as portions are exactly delimited. These requisites characterize the contract defined by the jurist no less than the one from which science is born. From this similarity we get the double use of the terms *attribute* and *property*.

Geometry in the Greek manner goes back to the Egyptian *Maat*. This word signifies truth, law, ethics, measure, and portion, the order that comes out of disordered mixture, a certain balance of justness and justice, the smooth rectitude of a plane. If some Egyptian chronicler, and not Herodotus, had written this story, we would have concluded that this was the birth of law, for it is as if a single process of the emergence of order, which the Egyptians had been orienting toward legal proceedings, had been drawn toward science by the Greeks.

Law precedes science and perhaps engenders it; or rather, a common origin, abstract and sacred, joins them. Beforehand, only the deluge is imaginable, the great primal or recursive rising of waters, the chaos that mixes the things of the world—causes, forms, attributions—and that confounds subjects.

This sounds just like the problems we have today.

Thus was concluded a social contract—will we ever be sure of this?—out of which politics and laws were born. This contract may be a mythic or abstract notion or event, but it is fundamental and indispensable to understanding how the obligations that bind us to one another were born, assuming that we don't want to see them as born of original sin or of our very nature. The cord of the contract came before the cord of obligations. It is said that the social contract formed all traditional societies, including the one we're living in.

A second contract founded a totally new society, which must have been born in Greece, five centuries before Christ, or still earlier in the Nile valley, and which brought together some people whose qualities we can't really define—priests, officials, jurists?—by requiring them to bow to the necessity of exact measure and then of proof. All sciences followed from this contract, just as societies were born of the first one.

As long as it concerned only mathematics, the second pact didn't differ much from the first, for it was simply an agreement in which a common decision could give birth performatively, by being stated, to the matter in question. As soon as we both want this to be my property and that to be yours, it is immediately so. In mathematics, the contract goes only slightly further; we have to agree on the properties of a statement or a geometric figure, and whereas a statement can indeed depend on our decision alone, a figure behaves like an object independent of us. This need for agreement about properties is behind Socrates' real-time demand to his partners in dialogue, which we find so tiresome, concerning every single word and thing. In fact, he's requiring that they continually sign this sort of contract, which underlies, in minutest detail, Plato's dialogues.

The scientific society, or the philosophic society of old, is born of these interminable signatures, without which no debate could take place. But, on the other hand, this society can only be born in opposition to traditional society, as if the new contract didn't use the same terms as the old. The connection that obligates us can go beyond us, as do the figure and its properties. As a result, almost all the signatories of the scientific pact, we shall see, appear before courts founded by the social contract, claiming that the old pact has no jurisdiction in the new decisions. There exists another world, that of mathematics for example, which goes beyond the performative domain.

The proof by contradiction, or *reductio ad absurdum,* the first proof that can be said to be conclusive, unfolds like an adversarial trial, in which, prior to the judgment, a certain thing belongs both to a set and to its complement: a decision must be reached. The adjective *apagogic,* which describes this primal proof, also comes from a legal verb, meaning to arrest a wrongdoer or pay off a fine. But in this case the deciding authority is beyond our control; the number imposes on us its law.

Whence the enormous belatedness of physics with respect to mathematics: it is infinitely harder to agree on a fact than on a statement or on a figure that we have at least constructed; it's even less easy to agree on the agreement of a fact with a statement.

At that point the contract will introduce a third authority: the world. Already, physics gives an idea of the natural contract.

GREECE'S WAY. Standing before the pyramids of Egypt, facing the sun, Thales is said to have invented the famous theorem of proportionality: enormous Cheops is to its shadow as less gigantic Chephren and Mycerinus are to theirs, and even as my ordinary-sized body and that little stake in the ground are to the dark streaks they project. Immense tails for the huge tombs, a trace for the minuscule stake, of course, but the proportion of the mark on the sand to the things themselves is preserved for whatever and whomever, just as a scale can balance two weights, one heavier and one lighter, by varying the length of its beams.

Here, then, is the oldest, pre-Aristotelian definition of distributive justice. To each according to his size and capacity; what is different turns out to be the same, the form remains stable for every size. Each drinks his full measure, whatever the size of his glass: every being has what it is. Better than the old measurement of crisscrossed fields, this absolute measure exhausts and includes all relativity, or discovers an invariance for all variations.

Does Thales turn *Maat,* Egypt's measure and justice, into geometry? Socrates shouts at Gorgias: "You want to prevail over others because you don't know geometry!" Indeed, knowledge of proportional equality proves, for all to see, that in the light of the same sun we display shadows proportional to our sizes. The world as such writes similitude upon itself, like a natural form of justice. How, then, can anyone claim superiority?

That which is written under the sun, in lines and shapes on the sand, and which compels general agreement through proof, must have soon passed for natural law. Hitherto unknown, absent from all human archives, not written down by the scribe's reed pen, this law was automatically projected beneath the sundial's gnomon, as beneath the pyramids, at every hour of day and night. What is natural law? Geometry: it falls from heaven.

From its origins, the question of justice marches in step with that of science.

ALGEBRA'S WAY. In the last decade of the fifteenth century, François Viète was recognized by all his contemporaries as the father of the new algebra, distinct from the algorithms and calculating methods that were standard in the Middle Ages. Its inventor, a high-ranking civil servant and a sober specialist in Greek mathematics, called it *Specious Arithmetick,* from the Latin *species,* which we translate by *type.*

In his *Treatise on Algebra,* which appeared toward the end of the seventeenth century, John Wallis, an English algebraist and mathematical analyst, says, citing others, that the *Specious Arithmetick* of Viète, who was himself a jurist, has its origin in a custom practiced by specialists of Roman and civil law, who would write Titius or Gaïus, John or Peter, say, or, in other words, so-and-so, A or B or C, to designate the subject of a special case and describe him more easily.

Roman law thus uses a particular first name to set forth a somewhat general situation; in the same manner algebra uses, in place of numbers, letters, whose values are less specific than those of arithmetic's numbers, but which vary within foreseeable and defined limits. It's a type, in the ordinary sense, or species, in the usual legal sense: a knowable singularity, if you will, a formal and concrete individual, manipulable as an index, more unknown than known. In concrete cases of guilt brought before the courts, Titius must be made into an individual. This is a different abstraction from the universal one used by geometers.

The route from the equation to its solution mimics the path from the legal text to the judgment, from jurisdiction to jurisprudence. *Specious Arithmetick* resembles a work of casuistry, in the sense of a general description of particular cases. In the end, at the conclusion of the proceedings, x equals 45 just as Titia equals Anne.

A moment ago, law preceded geometry; now here it is at algebra's origins. Does this precession never end in the sciences? Can it be generalized to all knowledge?

THE BIBLE'S WAY. In the beginning, God speaks the law, plants the garden between two seas, displays every tree pleasant to the sight and good for food, amid which are all the beasts of the field and the fowl of the air, and finally dictates to the first man his conduct: of this thou mayest eat, of that thou shalt not eat.

But Adam disobeys, and nudity, unhappy self-consciousness, exclusion, wandering, labor, pain, and pain in childbirth punish him from generation to generation down to ourselves. Our story and his tears can be explained by an ancient trial: before there was original sin there was law and a legislator, and from them came the sentence and everything that followed.

Was it really about eating? Of course, but not in the way one eats from hunger, since in the bosom of paradise everything was offered in abundance, but rather in the way one goes on snacking after being sated. Desire is born beyond need, after the calming of the body and the senses. If you eat of this fruit, you shall have knowledge, which sheds light on the question of evil, and you shall be as God. It's about comparison and knowledge.

A single word makes clear the nature of knowledge in its very foundations and origins: knowing comes from imitating. You shall know, you shall be as God. To be sure, knowledge always makes comparisons to a model, in this case a sublime and absolute one, but more important, knowledge, dynamic and all-consuming, which runs from the rearing of the human child to the glory or misery of the elder, only begins and develops because it is driven by the inextinguishable fires of imitation. Now imitation brings us to both good and evil, just as Aesop said our language does: no one learns without imitating, but it leads to mortal jealousy. Thus the knowledge of good and evil is identified with knowledge or science itself, which proceeds from the desire to be and to act as God; this divine desire leads irresistibly to evil. The questions of ethics and law that we have been asking about our own effective and competitive knowledge are already implied in this primal scene. To imitate, and thus to dominate; to prevail, and thus to destroy.

Law and science face off: the commands of law and the desire to know.

How can it be that the knowledge of good and evil will come from the fruit of the tree of knowledge, when the Devil and the

Good Lord were already opposed, through the woman and the man, as the very names of good and of evil? What else is there to learn?

In the garden between two shores, among the proffered fruits and the peaceable beasts of the field, the *libido sentiendi,* a dream of love and feasts, happy, modest, silent, and scorned, serves as a backdrop, a source, and surely also as an excuse for the transhistorical confrontation of two other libidos: the *libido sciendi* and the *libido dominandi.* The *libido sciendi* is a craving for knowledge so much more powerful than the *libido sentiendi* that the whole human race doesn't hesitate to set aside, in favor of its curiosity, and for all time to come, the entire Edenic satisfaction of the senses, though that gratification was within reach of the body. The *libido dominandi* is a never ending will to dominate, the most devastating of the three, the incontestable mistress of universal history. The senses are thrown to the winds so that brains and dominance can fight it out at their leisure.

What if the three characters of the primal scene simply embodied the three libidos? God, power; the male, knowledge; and the female, pleasure? The latter is pushed aside so that the first two can fight as much as they want, the way two billy goats pretend to want the same nanny, the better to satisfy their true passion, that of dismal and monotonous domination.

Do not eat of that fruit, the product of arborescent knowledge. The master lawgiver dictates the sensory conduct of the person who wants to know. And the tempter, going beyond the exquisite delights of the taste buds, beyond the ecstacy of knowledge, immediately pushes man toward domination: you shall be as God. Like him, you will organize and legislate. Imitation quickly passes over the *libido sciendi,* learning, to aim directly at the *dominandi,* the ambition of power and glory. Sensual desire, in its femininity, is out of the question here, and quickly takes leave of these mockeries: the body makes its requests in hormonal innocence. Amid a candid and yet already deceptive display of delights, the first trial opposes and binds together the will to power and the will to knowledge, law and science.

Being a lord almighty, God gives and speaks the law, which is so performative that it creates or carries out what it says when it says it. *Fiat!* The world itself is born of this command. Female and male together, in submission, weak, trembling with desire, they seek to know, and thus put at risk, through imitation, their peace, calm, plenty, and innocence, and their posterity. What utter, raving madness to give up what was certain in order to bet on the unlikely, and all for only a hope. I call this hallucinatory insanity prophetic, since it announces all of time to come. Here is the first of the meetings that confront kings and prophets, throughout history sacred and profane.

In the beginning, knowledge contests law and comes into conflict with it. Law wins, of course, since science is still faulty or sinful, but science produces the wandering of history, the drift of time. The beginning of the sciences engenders universal history.

What is knowledge, what is science? The set of deviations from law and its stable equilibrium, the oblique worries that lead to every evolution. I think, I weigh, I distance myself from law, which I have no right to do.

This, at least, is the tale of the singular emergence of Judaic, and thus Christian, culture, in which knowledge asserts the right to contest law. It has so contested it that it has killed it. The death of God amounts to that of the legislator.

Does our contemporary question reverse the originary one? What right will our rights of contesting our knowledge conquer?

Our Roots

Gaïus says somewhere that every obligation is born of a contract or a fault. If, in the term *obligation,* we read a bond that brings together or that subjects, we shouldn't hesitate to see, at the origin of the term *contract,* a similar cord that draws or tracts us together. Thus the theory of the social contract only repeats, tautologically, the necessity of collective connections: bond to bond. Moreover, Gaïus proves its equivalence to the theory of original sin.

Roman law secularizes this primal fault. Once again, metaphysics or formal discourses correspond to myths: analyze the state of nature or recount the wonders of the first garden and only the mode of expression differs, not the meaning. Obligation implies either fault or contract, it doesn't matter which.

Here, therefore, as is often the case, a story is as good as philosophy.

We're implanted in cultural soil, which was once identified with the natural earth, by at least four long roots: Greece, Israel, Rome, and Egypt. The use of scientific languages awakens the memory of the still-living thought of the ancient Greeks who defined them; the vague feeling of following the flow of history brings back the memory of the prophets of Israel, whose writing plunged us into it; we tend to forget our Roman birth; and almost from the beginning we lost track of our distant Egyptian origin.

Stuck in the past up to our waists and in the primal sometimes up to our eyeballs, even as we fly free and unfettered above the atmosphere, we have Semitic and Indo-European traditions coursing through the bluish blood vessels and the pale nerves of our legs, and variously tangled mixtures of these traditions reach our heads and mouths.

Every state since ancient Egypt and Rome has tried to emulate the durability of their immense and stable empires, by far the most long-lasting of all Western history. It is impossible even to imagine rivals to them in this regard: immobile and eternal stone statues, Egypt and Rome are and remain beings of law. The Twelve Tables sustained Rome and the *Maat,* Egypt. It's not enough to conquer; one still has to administer, and the strongest is succeeded by the most just. Pure law remains Rome's invention. It reduces both the mythic and the speculative to its own abstraction.

We think we've lost the memory of a form of organization that in fact still surrounds us, because it's easy to forget what endures, and only the superficial agitation of what changes awakens us and delves into our memories. Most of our references lie in darkness. Egypt and Rome did not produce much knowledge. What they had they did not develop, and thus they remain in obscurity. There, law prevails over science, which is why law preceded geometry and algebra, the sequence mentioned earlier.

In contrast, Athens and Jerusalem are beings of knowledge, small city-states without the possibility of becoming empires; they are torn, chaotic, more often beside themselves than calm inside their walls. Although they remain bound to law, especially moral and religious law in the case of Jerusalem, they spend their history contesting the right of law.

Prometheus, whose name signifies primordial learning, the origin or archaeology of our knowledge, never stops dying, tied down to a rock on Mount Caucasus. Similarly condemned, Socrates, the teacher of teachers, drinks hemlock, a spectacle for his admiring disciples. To promote its own laws, every form of knowledge goes to trial: in one case against Zeus, king of the gods; in the other against the city's judges or archons.

Adam, the originally reproduced man, invents history and throws himself into it alongside Eve by staking paradise against knowledge, by contesting the first divinely spoken law. He heralds, I would say, the encounters between prophets and kings, the perpetual trial, the motor of historical process. The trial of Jesus reprises and renews this tradition, fulfilling it and splitting it into two paths. A certain kingdom is not of this world, this world of law.

In each of these cases what matters is not particular legal actions but the fundamental trial that shakes the foundations of legality. A choice must be made between law and knowledge, for knowledge begins at the same time as the question: what is justice? Whereas in Egypt or Rome, only justice has the right to ask questions, the question "what is justice?" is the first one asked by Jerusalem and Athens. Because they forsake asking this question, neither Rome nor Egypt really produces a body of knowledge; conversely, by asking it, Athens and Jerusalem forsake the kingdoms of the earth. Science prevails over law. Greek sages, Herodotus and Thales, had to travel through Egypt, and Viète, a Christian jurist, had to tear himself away from Roman law.

The contemporary debate that opposes, sometimes violently, these two authorities—science and law, rational reason and prudent judgment—has deeply moved our flesh and our word since the beginning of our history. The history of our knowledge follows the temporal trajectory set in motion by this trial, still vigorous today, an originary source and a perpetual motor.

General History of Trials

The two founding heroes of chemistry and mechanics, Lavoisier and Galileo, who were charged before the respective tribunals of the Revolution and the Church, and thus celebrated in the judgment of history, cast shame on the justice of their time. Yet the Earth does move!—everyone hears scientific truth finally burst forth in the face of absurdity—the Republic has no need of scientists! Law once prevailed over science; science now wins against law.

Who today doubts this clear division between light and darkness, this decisive verdict in favor of science, this striking reversal of what the bench of old had ruled? But who suspects that to accept this verdict is not to champion the cause of a defendant or victim, as might be believed, but to become part of the jury in a new trial? The magistrates of 1794 and the cardinals condemned scientists; we, in turn, condemn the revolutionaries and the churchmen: in formal terms, what has changed? Real or virtual, a tribunal thus sits in permanent session; the trial goes on. Can't truth do without judgment?

Law made pronouncements on science: on the basis of what knowledge? Science decides on law: by what right?

Thus neither Galileo nor Lavoisier can or should be seen as an exception, for we find judgments and trials in abundance throughout history. From the beginnings of scientific knowledge, so hard to pinpoint, the first dialecticians, astronomers, or physicists appear before the tribunals of the Greek city-states, on counts similar to those faced by modern scientists. They come through badly or not at all.

The sciences begin in such legal actions; they enter history through the courtroom door. This should be no surprise; when science goes before the tribunals, a synthesis is already achieved between its two histories, internal and external. The internal history requires a judgment of truth, a decision about whether Anaxagoras and Galileo and Lysenko are wrong or right; this is a point on which even Anaxagoras and Galileo seek assurance. The external history has the sciences join or develop into schools or pressure groups, and it demands that their truth be socially canonized. Individuals or associations appear before a given court, and fragile truth is thereby reinforced, for the decision handed down casts it into an officially sanctioned time. In the final analysis, there is no general history of science that is not judicially recorded. No science without trials; no truth without judgments, whether internal or external to knowledge. Its history can't do without tribunals.

Science never again leaves the courtroom. Michelet saw quite well that sorcery or witchcraft trials, far from testifying to the absurd cruelties of dark eras, express time and again the inevitable, fundamental, inescapably ritualized encounter between knowledge and law. Knowledge is always obscure and nocturnal before reaching the light of day, always sylvan prior to being given voice in the city's public square, whereas law is always clear and distinct until knowledge, in its turn, drives it back into the shadows of ignorance. Yes, every natural science, whether of the sorcerer or the sorcerer's apprentice—in any case taking no responsibility for social affairs—every science indulged for a while, or is still indulging, in the witch's sabbath. It cuts circular furrows in the grass, it tears a round hole in the ozone layer, it exposes the world to great dangers...

In this respect, Michelet, before Bergson, and Bergson, before our contemporaries, both succeed in sketching the inside and outside of societies, the worldly world and the other—for instance, worldwide—world, but of the two only the romantic historian senses the importance of the tribunal as a place of contact or recording, a sieve, ticket window, or semiconductor between the two worlds. There is only one trial, of a single witch, there is one case and a single scientist, and through this exemplary legal action the history of our knowledge and its multiple bifurcations is continuously decided.

Trials, Continued

ZENO OF ELEA. He shook up the Greek philosophers of his day, diabolically skillful though they were; the mathematicians of classical calculus; and modern logicians, although new methods have given the latter an apparatus superior to his: who has ever been more ingenious than Zeno of Elea? Starting at the origins of mathematics, he managed to give the most abstract thinkers something to ponder for at least five times five hundred years. Who could be more ingenious than he, the inventor of dichotomy, the division of an itinerary into two parts, then a new division into two segments of the part still to be covered, and so on infinitely, so that the traveler never reaches his goal and the thinker begins to conceive what is called abstraction?

He had been given, it seems, the epithet "Amphoteroglossos," which means he was accused of having a ready and forked tongue, like a viper's, to speak pro and con, yea and nay, white and black, false and true, with equal plausibility and rigor. In fact, he invented dialectics, the legalistic art of winning in dialogue or of confounding one's adversary through interrogation. This method was doubtless borrowed from him by Socrates and by all those for whom truth is defined by defeating someone else: this is obligatory conduct in judicial debate, and it led Zeno inexorably into court.

Diogenes Laertius relates that Heraclides Ponticus reports—all of which means that I'm telling this without knowing what's true and what's false in the string of criticism concerning these tales, which have been lost and then found in such a fragmentary memory that, like Zeno, you can always find an intermediate point between where you come in and the goal you're trying to reach—reports, in any case, that Zeno was arrested for plotting against a tyrant, whose name varies according to the source, if it's mentioned at all. How paradoxical that this testimony manages to reach us, when everything suggests that forgetfulness should win out! In short, here he is on trial, the very creator of the most fearsome weapon in judicial jousting.

Name your accomplices, the king orders. Your guards, answers Zeno, your friends, the whole palace entourage. An appalling strategy on the part of Amphoteroglossos, who shrewdly isolates the holder of power from all those whom he believes love him. Moreover, this lie frees the city, since the tyranny, wasting no time in putting to death its own supporters and immediate protectors, is so weakened that it falls. It's a victory of knowledge, invented by the philosopher, and a victory of courtroom form over the man who instigates, dominates, and organizes the trial. A triumph of dichotomy, whose scalpel cuts decisively through all bonds, even human ones. An analytical success.

But suddenly Zeno of Elea declares that he has confidential revelations to make, which he can and must give only by whispering to the proper authority. Unfettered, he approaches the tyrant, the only one entitled to hear what he has to reveal, and his mouth touches his ear. No, he's not speaking, but attacking and biting. His jaws clenched, a leech, a vampire, a tick, the inventor of dialectics wouldn't let go until he was dead. You can hear the royal cries of pain filling the courtroom. Socrates calls himself a gadfly in his *Apology*, and says that his fellow citizens will have no peace from his stings and bites until he breathes his last. Is it possible to imagine a living organism, a horse, stag, or passerby, who wouldn't be driven crazy and wouldn't try to get rid of this piddling insect by crushing it? Tearing the leech from its hold on the skin, killing the parasite?

Does knowledge really parasitize the law, with which it chooses to maintain a relationship? But of course: it imitates it, mimics it, theorizes its form, refines it, and finally fights it to the death—its own or the judges'. The whole history of the Greek beginning of the sciences tells of the shared and tragically eventful life of these enemy twin sisters, justice and justness (or accuracy), reason that judges and reason that proves. Today's question is: when and how do they become symbionts?

One of the first sciences common to almost all those who have lately been called Presocratics, logic, entailed a formalization of judicial debate. Both logic and the arts of language came from the courtroom, from the various tribunals, in other words from the relationship between well-conducted arguments and death. All the refinements of rigor—contradiction, proof, *reductio ad absurdum*—come from being experienced and tested less with respect to external or natural facts than with respect to human law, which is infinitely more present and dangerous than natural facts.

From the tragic comes the judicial; from the judicial, logic; and from these three *logoi*, the scientific *logos*. It had been some time since water, fire, or fierce beasts—nature, in other words—had put the ancient Greeks in danger, whereas death stalked them in public assemblies. The dialectic and logic taught for a small fortune by the sophists answered the need to defend oneself in debates whose outcome, sometimes, led to exile or death.

A variant. Another story tells that Zeno cut off his own tongue and spit it in the tyrant's face. No more tongue to speak, no more ear to hear: the message or debate, rhetoric or dialectic, can no more pass through the space of the courtroom than Achilles, the arrow, or the tortoise can cross the interval that separates them from their goal. What parasite, in the sense of static or noise, intercepts the transmission of the message? But since Amphoteroglossos has a forked tongue, which segment did he cut off to throw in the face of power? He still has another left, so he can go on speaking!

It is said that the citizens, seeing this, were beside themselves with rage and together stoned the tyrant. This time the rocks reached their target.

If Zeno invented dialectics, he succeeded in giving debate, interrogation, and all the conventions of trials their canonical form. If he puts a stop to the circulation of messages, by cutting up tongue and ear, sending and receiving, with his teeth, he destroys the possibility of any proceeding, any debate, any representation, thus any contract, thus the collectivity's foundation. Then the judicial, undone, slips back below its conditions, toward the origin, the sacrifice, the tragic. Just as tragedy precedes every authority, every proceeding, so the death penalty follows lynching.

The foregoing analysis leaves a residue: there's still a mouth and a tongue, the tyrant's, to cry out in pain, and an ear, Zeno's, deaf to these cries. But for the texts of our sources, the circulation of messages in this direction doesn't count. The philosopher speaks, not the king; the king listens, not the philosopher. Guess who parasitizes whom; figure out from there who comes out ahead. Science prevails over law.

THE APOLOGY OF ANAXAGORAS OF CLAZOMENAE. "Doesn't your country interest you?" one of his contemporaries asked the philosopher Anaxagoras, whom he saw living apart, in solitude, attentive to celestial doings. "You couldn't have put it better," he answered, pointing upward, "it's my only concern." In other words: my kingdom is not of this world, worldly, but of the other, worldwide. Do we live within the walls of our cities or under the starry dome? In which of the two? Do we dwell in one more than the other?

Anaxagoras contrasts the sciences of nature with those of the city, while bearing witness to a stable time when everyone was concerned only with city sciences. The social sciences put astronomy on trial. By what right?

Just a moment. During his passion, Jesus Christ, too, speaks of another world, different from this one, where the tribunal that's judging him has no standing. He calls it a kingdom. Wherever a king reigns, politics and law exist, and thus tribunals, just as they do here below. In fact, everything will end with the Last Judgment, after the end of history, when today's victim will return to sit at the right hand of the Father and to judge, in his turn, the living and the dead. The last tribunal of the other world resembles, in its form, the first tribunal, in this world, for in both we find the customary appeal to a supreme authority, the last, which is without appeal. At least the world beyond follows law.

In the court cases brought against science at its birth, the same appeal is heard, but on a totally different level. Yes, Galileo and Anaxagoras lodge an appeal, but it is to the earth that *does* move, or to the sky, the philosopher's country. These worlds are not kingdoms equipped with tribunals, but rather places outside law, without politics or kings. There it is, gentlemen of the jury, nature! Land without rules, truth without judgment, thing without cause, object without subject, law without king. Is science's historic task to invent a new form of justice in this land without a contract?

The question of country put to the philosopher-physicist requires more from him than you might think, for it criticizes him and attacks him mortally. What! you don't give a damn about any political and social commitment? You don't read the newspapers, you don't say your morning prayer? It sounds like Sartre or the political moralizers who came before and after him. And nobody dared to reply to those terrorists that they were ignorant of physics! Ancient Greece sometimes gave the name philosopher to the hero who resisted to the death the obligation to be political, whereas Sartre demanded conformity to the political so as to seem a philosopher. In the age of my fathers and their successors, the sages had taken (and still hold) the place of public prosecutor, demanding condemnation in the name of the city's dominant forces. The bastards!

By what right, then, does some citizen criticize Anaxagoras? By this fundamental legal right that founds the existence of the city and that is sometimes called the social contract. If, because you're observing the planets, you lose interest in your country, then you break the contract that unites us, and thus society must logically exclude you, condemn you at least to exile and at most to death. The conclusion is rigorous, in both senses of the adjective.

This conclusion supposes, indeed, that the social contract concerns everyone, without exception. How can the general will be defined, except as the will of everyone, and not as that of everyone minus a few, such as Anaxagoras and scientists? If you're not concerned with the city's affairs, you're excluding yourself from it by your own choice; because you remove yourself from the general will, you pronounce your own sentence. Like the contract, this trial can remain virtual, but it can also become real at any time. As can the death sentence. The contract, in its logic, knows no mercy.

What's the meaning of this fine totalizing of the group's composition and of everyone's activities, without gap or exception? It means this, which is a lot: that the virtuous citizen's knowledge and his constant occupation consist of knowing in real time what the other citizens are doing, and of making it his own business. Everybody knows everything about everyone, and everybody is busy with everything that everyone is thinking, saying, and doing. This is absolute knowledge, or rather absolute information, total commitment, a contractual obligation or a complete system of cords and chains, the total transparency that is the aim of those who write and read newspapers, whether they're printed or spoken or in images: this is the ideal of the social sciences. Hegel was only off by a little: the philosopher reading the newspaper is indeed praying, but to the idol of absolute information: nothing, in principle, escapes him. This universality founded the city-state of antiquity and expresses its ideal. Those, like Rousseau, who describe it as something to be regretted, are concealing, or are unaware of, the colossal price at which it is purchased. Let's make a distinction, by the way: the information given by social science remains banal, for it repeats what everybody knows about everybody; the information given by natural science, on the other hand, can be calculated and is proportional to rarity, and we call it knowledge.

When everyone knows everything right now about everybody and lives by this knowledge, you have antiquity's notion of freedom and the ideal city, and also the ideal of modern philosophers since Rousseau, the ideal of the media and social science, of the police and bureaucracy: poll, clarify, inform, make known, expose, report. A terrifying nightmare, one that, if you've lived in small villages or large tribes, you'll want to avoid all your life, for it is the height of enslavement. Freedom begins with the ignorance I have and wish to preserve of the activities and thoughts of my neighbors, and with the relative indifference that I hope they harbor for mine, for want of information. Our life in enormous metropolises makes us dream, as if of a lost paradise, of these appalling Athenses where continuous and total information made everybody the slave of everyone else. An astronomer, Anaxagoras, or any other physicist, conquers freedom in nature's space.

The city-state of antiquity knew no police. It needed none, since everybody's information sufficed to monitor everyone's conduct in real time. That citizen whose virtue was celebrated throughout history, from Plutarch to the French Revolution, would appear to us, if he were brought to life at our side, to be nothing less than a full-time tattletale or snitch, an insufferable character, an informer or reporter, always running around telling everybody whatever can be found out about everyone. This absolute and totalitarian informedness, controlling and dangerous, belongs nowadays, in theory, to the police chief. Contrary to tradition, I no longer praise the ancient city, I condemn it, condemn the policeman's absence, which shows that everybody took on the tasks of surveillance and repression. Where there are police, great, there's some chance of freedom.

Athens remained ignorant of the role and official function of a public prosecutor. Each citizen could fill this office, and, in the public interest, accuse another before the court. This is yet further proof that everybody played the part of spy and inquisitor for everyone else. Contemporary thought inherited from this. How few philosophers, indeed, in the last half-century, haven't seized and enjoyed the role of state's attorney, prosecutor, accuser; they've denounced abuses, crimes, errors, hypocrisies, as if they were journalists: this is their rightful place. No, our philosophy shouldn't be called that of suspicion, but that of denunciation. But by what right does it give itself this job? Isn't it ever wrong, doesn't it ever make a mistake? In the city-state of antiquity, everyone enjoyed this right.

When an organ appears, in the course of evolution, it frees the organism as a whole from the crushing weight of the function it carries out. Better to have the policeman and the prison, these highly visible, specialized organs, recognizable by the uniform and the bars, than the omnipresent eyes and ears of one's associates and of those all-seeing strangers who represent the virtual contract and act on its behalf. Our liberty is defined in opposition to this monstrous ideal, and it depends on a degree of ignorance, on missing information. The liberty of the modern era reverses that of antiquity by freeing us from the crushing weight of that absolute and global information, henceforth either useless or preserved in data-processing records. We don't know how lucky we are that our minds are relieved of this social cord: as a result, they can turn to real sciences!

Here, once again, we seize a real-life possible origin of scientific knowledge, in relation to the social contract. We doubtless learned or invented sciences in inverse proportion to the ancient mode of being informed: the less we busy ourselves with others, the better we like them; the less we gossip, the more we know the world. The less we know of what's banal, the better we grasp what's rare. The methods and ends of the social sciences are only those of policing; their content is mere information and their history purely archaic. The empty place left by collective petty trivia has been filled by knowledge, a modern creation. This is one of the lessons of Anaxagoras, leaving his old country for the new.

Now let's suppose that the ideal of total social familiarity is realized, the way Athens, to some extent, and Sparta, doubtless much more, once knew it: we can immediately understand how each of these once virtuous citizens, unlike us, might find it monstrous that a single one of them could abandon such knowledge and such activity, for he would be destroying, by this very act and on his own, that universality. If somebody stops knowing and saying everything about everyone else, not only does he leave the general will, he destroys it. Take just one planet away from the solar contract, and this change threatens the movement and the stability of the whole system, at each and every point, for it can only keep this equilibrium and these orbits by remaining as is. Imagine a perfect system; it turns out to be the most fragile system possible, and it must maintain its universal law, the same throughout. Conversely, for it to adapt to changes, it must be conceived and built with play, in the sense that cogs have play, or give. Any evolution can only be born of fragility. Our modern contract of freedom thus demands some ignorance: I don't know what my neighbor says and does, and I don't report any of it if I do happen to find out, unless I claim to be a social scientist or I sign up as a police informant. And I count on my neighbor's thinking and acting the same way toward me. As a result, the contemporary contract partially reverses Rousseau's ancient-style one, written or not. We make up a limited-liability corporation, and our freedom resides in this limitation. It comes in part from spaces of non-law, where nature can get through.

So, then, Anaxagoras studied the sun and the moon, the Earth and the formation of the whole, the Milky Way, the world's movement, for nature was more interesting than public affairs to this physicist—a physicist in the very old sense.

Let's turn for a moment to the famous trial of Socrates. Meletus accuses him, before the court staged in the *Apology,* of being likewise more devoted to physics than sociology, to use our terms. Socrates denies responsibility and counters by accusing Anaxagoras of this misdeed: "So go spend a drachma on his books," he says, "where you can read that the sun is a stone and the moon is earth" (*26d–e*)—as for me, Socrates, I've never said anything of the sort. As usual, Socrates replaces the interrogation he was supposed to undergo by an interrogation that he conducts: the very person Plato casts in the role of victim now becomes the inquisitor, and the trial of the physicist Anaxagoras can be discerned, like a play within a play, in the trial described in the *Apology*. Even before a court in session, the permanent Socratic tribunal never recesses; it is still more inescapable than the one that will condemn him. Even on the defendant's stand, Socrates can't stop accusing. He's a permanent public prosecutor, carrying his portable tribunal on his shoulders in the streets and public squares, and thus plunged up to his neck in the absolute information required by social science. Right in the middle of his trial, an ordinary trial since the question is only whether or not he broke the laws of the city, Socrates opens the conditional, transcendental trial of the man who removes himself from the city and its law, a case so fundamental that it irresistibly breaks through the Socratic plea, which Plato, nonetheless, makes into a foundational discourse in the *Apology*. The juridical gadfly couldn't care less about the moon!

Accused of having claimed that the sun is burning, Anaxagoras was condemned; that goes without saying. On leaving the courtroom, did he cry, "Yet it *does* burn!"? In fact, he likened it to an incandescent stone, bigger than the Peloponnisos; from this stone fell the brown, chariot-sized meteorite, whose impact, around Aegospotami, assured his fame, because he had foreseen it: how can a thing like that be predicted?

A big chunk of nature falls in the middle of the city; a fine physical-science object suddenly drops into social science's terrain! Terror in the city as in the fields; not, as one might think, because of the exceptional miracle that came inexplicably from the heavens, but because the worldwide world is revealing itself to those who know only the worldly habitat. This is what's rare. This is the miracle, to tell the truth: that nature manages to break through culture's fast enclosure. The stone falls from the firmament into the city, from physics into law; the trial of Anaxagoras falls into that of Socrates. Stupefaction: the real miracle is falling bodies. Nobody had thought to supply a god for gravity.

Immediately, the social sciences take over: the body is not a body, nor is the inanimate inanimate. The weighty mass becomes a god and the rock a statue. The worldwide event is quickly brought back inside the borders of the worldly; religion attributes to men what really came from the sky. The city's closure upon itself is sealed anew.

At this point the doxographical tradition is shaky; the sources are uncertain about whether to attribute the meteorite's prediction to the philosopher Anaxagoras or to the famous king Tantalus. Why this unexpected pairing?

Just as the trial of Anaxagoras emerges framed in that of Socrates, we can see the trial of Tantalus framed in that of Anaxagoras. The king was condemned to an eternal punishment: in the hellish caves described and popularized by Homer, the wretched man, tormented by thirst, exhausts himself trying to drink, while a cup approaches his mouth without ever reaching it. Likewise, racked by hunger, he can't eat. The agony of Tantalus is the image of our unsatisfied desires.

But in the Greek tragedians and in the poem of Lucretius, Tantalus, though still cast into hell, waits for a rock delicately balanced above to fall at any moment on his head, yet it doesn't fall. The tension of desire gives way to that of anguish, and we have a symmetrically opposite situation. Eternity adds up the differential instants of bleak terror and uncooled longing. Can anyone who suffers in real time from renewed appetites or constantly reawakened fears be thought of as dead? Certainly not, for such is the definition of life.

We all survive under the sun, exposed to the socially and humanly meaningless fall of a celestial fragment when it leaves the moving system or the nearly stable vortex that bears it. When will it fall? What's the use of predicting, since we are sure of death and only unsure of when it will come. The time of our death is absent, excused, from any absolute knowledge.

So, the stone falls on the city, the earth quakes and thus shakes our walls and our constructed certainties; nature bursts in on the citizen, who believes only in the assurances provided by human labor and by the political order or police (you have to admire the wisdom or madness of languages in which insurance contracts are called *policies*). We ought to admire the madness or wisdom of our ancestors the Gauls, who feared, it is told, that the sky would fall on their heads: indeed, that could happen this morning, unannounced, and what's more it will surely happen some fine morning. Their madness or wisdom is just like ours, alive and brief, the eternal anguish of the king in hell, threatened by the rock.

Question: where do you locate this hell? Right here, as far as I know, under the still constellations, under the incandescent stone of the Anaxagorean sun, in the worried time while these heavenly bodies remain delicately balanced, for the duration of our short life or of the hellish and mediocre history that stretches around it. Hell, considered as a thing apart, quite nicely defines the place of nature, understood as the space of exile and banishment: if the stone threatening Tantalus falls on his head, then it returns to its natural place.

We keep forgetting about meteors; we're always attributing human causality to thousands of events that are actually determined by climate. Our ancestors the Gauls, like myself, would have preferred geography, which is so serene, to history, so chaotic, and preferred Montesquieu to Rousseau. The latter must be taken literally when he says that, after the contract, there is no more nature except for the solitary dreamer, whom society has forgotten. Meteors vanish in political philosophies, which are every bit as a-cosmic as social science, after a few initial moments that are indeed evoked or thought of as originary, the better to eliminate the world.

Thus, even though Pericles at the height of his fame defended him, Anaxagoras found himself condemned to exile and banishment from the city for having said that the sun was nothing but a rock that could fall. But he had already been living outside politics. In the words of an unknown scholiast: "Having become a philosopher of nature, Tantalus was condemned for demonstrating that the sun was on fire. His punishment was to be exposed to its rays until he was immobilized by the shock of its burning." Doxography is sometimes coherent in its historiated chaos; it tells us that hell is no different from the worldwide space under the sun. So Tantalus was cast out.

What is nature? The city's or culture's hell. The place where the banished king was cast out, the city's outskirts, suburbs. This exclusion shows that the distinction between two spaces or worlds, worldwide and worldly, nature and culture, presupposes a judicial decision. This is no customary or routine decision drawn from jurisprudence but an extraordinary judgment handed down by a fundamental tribunal in the course of an original or transcendental trial: a First Judgment, as in Last Judgment, pronounced by this tribunal, seated on the boundary between the two spaces.

Of this hell, Anaxagoras said, "The descent to Hades is much the same from whatever place we start." Whether you leave from Sparta or San Francisco, dying is done the same way. Whether you're exiled from Paris or Pisa, the world outside lies under the same imperishable sky. Twenty cities and one external world, under the sun, the same for all those excluded. A hundred sets of laws and a single desert exile; all suburbs look alike. A thousand cultures, one nature. A hundred obsessions, one way to breathe. A hundred thousand social science books presenting millions of pieces of information; one knowledge and rare thought.

A multiplicity of diverse lives and mournful grimaces for a single kind of death. Where do we get the universal? From demise. From expulsion. From the outdoors. From the hell of falling stones. Yes, from burning stars. From the other world. From a world without men.

Anaxagoras was condemned to death in absentia, as were his children, another source specifies. "Easy does it," he exclaimed, "nature had already promised us to death from the moment of our birth." Even if a hundred authorities hasten its coming, since they're incapable of delaying it—was the death penalty invented by this arrogant impotence?—a single authority, of last resort, one that we do not know, holds the power to decide the end of human life. So many untimely condemnations for one universal death.

As if they were unaware of their common fate, mortals are in the habit of assembling in communities to proclaim to one another that they've invented death. Common, then, if it is allowed to take its course, and only communal if the day of reckoning is hastened, death lies at the intersection of man-made laws and natural laws. In the same way, suburbs or the desert of exile, hell, outside, the space under the stars, all trace the spatial intersection of man-made verdicts and natural places. The tribunal and death arise on the same spot.

Who or what then condemns me to death? My body, my human

and living condition, the law of falling bodies if the sky drops on my head, the laws of combustion if I set myself on fire—or persecution by some tribunal? The penal code, the genetic code? Nature, or my culture? Their confrontation takes place before a tribunal, as if only the judicial could record the unicity of the laws of the world and of death in the face of the multiple and relative decisions of social codes. After all, Anaxagoras says that even nature condemns him to death, as if there were really a tribunal outside, and thus a legal order that can submit to its rules these two types of laws, those of natural science and those of social science. And thus law prevails over science.

The city casts out Anaxagoras, or else he dies, for having said—and the jury upheld the charge against him—that the sun is a burning stone. We live in exile, we die condemned; heavy bodies fall, including meteors coming loose from their orbits; fire burns and fills the universe with its heat. These are three natural laws that have been pronounced as such, canonized, before the tribunal of man-made laws. Law prevails over science, and the Greeks, mathematicians though they are, will not invent physics.

Except for brute force and the unveiling of glory that history offers, the only truth, in the beginning, is judicial.

Law never gives orders and rarely writes or speaks in the imperative; nor does it designate, that is, write or speak in the indicative. It uses the performative. This means that truth, the conformity of the spoken or the prescribed with the facts, ensues immediately from its prescription or its speaking. The performative makes speaking an efficacious act, a sort of fiat: in the beginning of the world, God the creator uses this kind of performative utterance: he speaks and things happen in accordance with his word, as if the creation of the world had been conceived as a law. Thus the law does not err, cannot err. There is no judicial error; or rather, a tribunal can be mistaken about the facts that it is authorized to judge, but the law that it represents is not mistaken. The arbiter, infallible because performative, is always right. If he is wrong, he has given up arbitrating.

The social contract generalizes this law of truth, when Rousseau says that the general will can never err. Of course. If the contract founds society, politics, in turn, is founded on law, since the contract is its fundamental act. Convention, as the agreed-upon coming together of many men, is founded tautologically on convention, in the sense of a contractual and conventional agreement. And law, being performative, does not err; therefore the general will can never be mistaken. Rousseau succeeds in demonstrating this paradoxical but obvious fact on which the life of the ancient city-state depends: the conventional, being infallible, is always true. Antiquity knew only truth produced by convention and glory, what we would call today the media and management. The true, which we think should be founded on something other than an arbitrary convention, is founded, on the contrary, precisely on this convention. The arbitrary is infallible. This is a fundamental theorem of performative law, though it appears paradoxical: the absolute necessity, the organic obligation of arbitrating. Death or this theorem.

The history of our knowledge starts from this theorem, engages in ruses with it, fights it and acknowledges it, hates it but cannot do without it. What is science, knowledge, and even thought? The set of confrontations between all the other foundations of truth and this fundamental act of arbitrating. Every certainty, therefore, must present itself to be recorded and confirmed, to be canonized, before a tribunal.

Taxonomy of These Cases and Causes

TIME AND HISTORY. A trial always ends by giving a ruling, by concluding about the case; judges apply texts and jurisprudence in such a way that their sentences contribute to nourishing, in return, jurisprudence and the evolution of law. The tribunal's decision thus opens up a new time. Not time that passes and slips by, as if left to itself, but a time worthy of reports and documents: a history. Perhaps we have nothing but law to transform time into history, or to canonize the former in the latter. Better still, history does not so much unfold by trials as act as a permanent tribunal itself.

An event creates a bifurcation; conversely, a bifurcation counts as an event. Since the pronouncement that concludes a lawsuit makes a choice among various paths, it closes up possibilities in order to open a single one, like a semiconductor, a lock, or a ticket window. The series of this kind of trials produces the group or succession of bifurcations along which history flows, along which time passes so as to be thereby canonized. These are the judicial vertices or nodes distributed in the network traced by the history of science. Space, logic. Every decision, as the word indicates, defines a region of space, concrete or abstract. Every decision not only cuts a tillable plot out of the chaos left by the furious swelling of the river or war, and attributes this plot without usufruct to a given person, but also and especially, every decision delimits concepts and their properties analytically. The first jurist of Roman law, the inaugural logician or set theorist, was the augur who, before observing the auspices, marked out *templa* or zones with his ritual staff among the possible locations in the sky. Law describes what happens in any kind of space, whether real, material, formal, or linguistic: the discovery and the division of this original space are the very origin of law. Its language, not prescriptive but performative, by describing sites and attributes, that is, places and properties, makes them into what they are.

Time becomes canonical and is transformed into history because we ascribe it to these sites as soon as they exist. The jurist invents this type of abstraction. Not constraint, not morality or policing, but an analytic cartography: in this, law looks like protogeometry. As if the two forms of reason, scientific and juridical, started out by analyzing or carving up an existential and categorical Earth, fundamental, transcendental, archradical.

Examples. In the beginning is the religious. Suppose that a given social group practices given rites. At the slightest deviation, the collectivity reacts and corrects to restore the equilibrium of the norm; if the divergence grows, a choice must be made between orthodoxy and heresy. Whence the religion-religion conflict that only a trial can decide: Jesus before the chief priests and elders; Councils versus Reformers—Luther, Calvin, or Michael Servetus.

But a given religion sometimes promotes laws contrary to those of the king, or of whatever regime is in power. Whence the religion-politics conflict, also decided by a trial, that of prophets face to face with kings, or that of Jesus, once again, before Pontius Pilate, during which the Redeemer pronounced the canonical words: my kingdom is not of this world.

With every verdict, a space opens up and a time is born. Churches and sects are defined and thrive, each with its terrain and history. In the same way, the trial of Antigone against Creon defines in space and engenders in time a certain morality in relation to political power, or a private right in relation to public law.

One after another, the sciences are born, each seeking to mark its original limits and prerogatives (I was about to say its jurisdiction). We will surely never know how or where or by whose efforts they really began, but we cannot forget the legal actions that sanctioned their entry both into history and into truth, the process that could be called their canonization.

The sciences split off from politics; their terrain is distinguished from collective space, their contract differs from the social contract, their language is neither spoken nor written like public discourse, and the history of their truths is full of bifurcations. Thus each science has its trial against the tyrant or power: that of Zeno at the origin of mathematics, that of Anaxagoras at the failed beginning of physics, that of Galileo at its successful emergence, that of Lavoisier when chemistry started, and so many little anti-Darwinian cases when modern biology was launched.

The sciences are distinguished from religion: their text differs from sacred writing, their truths do not have the same references. This explains Galileo's trial, once again, for astronomy and mechanics, and also the cases that stir up biblical fundamentalists faced with the theory of evolution.

What do the sciences have to do with morality? Morality is written in the imperative and knowledge in the indicative, like law, but without being performative, as law is. Today local and national committees of medical ethics are forming, seeking to conceive laws as yet unwritten, like those unwritten laws of love to which Antigone referred. Similarly, we need a collective ethics in the face of the world's fragility.

These successive trials delimit the respective spaces of the sciences and their prerogatives, by distinguishing them from the other domains and types of truth, which had already been distinguished by other trials. This multiplicity of fields (religions, politics, moralities, sciences, and so on) defines pretty well what we call secularism, a global and pluralist concept quite close to this distributive justice. Saint Thomas Aquinas, who first introduced a man-made legal order independent of a universal divine legislation, invented the effective usage of secularism, though certainly not the concept.

We know societies in which everything is religious, others in which everything is political, and so forth: in them, each social fact tends to become a total social fact. The local invades the global and becomes totalitarian or fundamentalist. Justice and secularism reverse this tendency and struggle against it by assigning specific places and prerogatives. As a total social fact, politics dictates to biology its Lysenkovian or Michurinian truths; if religion becomes a total fact, it will impose its dogma on Bruno, Galileo, or Darwin's disciples. Whence the scandalous trials, which the history of science does not so much suffer from as emerge from.

But if, suddenly, the sciences, which benefit from the aura of sacrificial victims and from the justified triumph of their own type of reason in the time of history and the space of the whole Earth—if the sciences were to take over by becoming, in turn, a total social fact, and dictating their truths to ethics, to laws, to politics, to religions, to philosophies, then injustice would return, symmetrically, from the other side of meaning, space, and time, putting this secularism into peril once again. Will we see the start of some trial, still inconceivable as of now, some completely different new legal action? It sometimes happens that that which contributes to liberation turns around and becomes a power to enslave us.

Because of this, the series of canonization trials continues, irresistibly, right inside the sciences, once they are canonized. In other words, they separate from one another, distinguish among themselves, and thrive by instituting, as one of their fundamental features, a network of jurisdictions such that none of them judges itself competent outside its own terrain. This procedure, sometimes called falsification, in fact quite resembles agricultural property law, the competence of a tribunal, or some political or military partitioning. Historically fluid, the classification of the sciences takes the form of cartography.

So the history of science resembles, like a twin sister, the history of religions cited above, and our cycle is complete. The old religion-religion conflicts, with their courtrooms for heretics, their burning of sorcerers later extolled as saints—these conflicts are renewed in interminable science-science conflicts, settled by the permanent internal tribunals that regulate scientific life. Thus the history of science leaves behind it as many outcasts as does the history of religion: Boltzmann kills himself on an Adriatic beach, Abel dies forgotten in his prime, we are forever recalling once-despised precursors.

Philosophers used to dream of a science of sciences; we have finally awakened from this dream. Even epistemology does not exist, except as a redundant discourse, a form of advertising. If it were reborn in the form of an epistemodicy, who would be able to describe the relations between judgment and truth?

Thus the trials of Socrates, Jesus, Galileo . . . are not exceptions, not by a long shot. On the contrary, they bring out a law of our history: that the laws of the city, the institutions, the social, religious, and political order all agreed at some point to lose at their own game. The archons of Athens, the pontiffs and Pontius Pilate, the cardinals of the curia—our history consents to putting them in the pillory, where they are joined by the members of the revolutionary tribunal that had the chemist Lavoisier's head cut off; or those of the English jury that drove the logician Turing to suicide, even though his inventions in computer science had contributed decisively to preserving the British Isles from Nazi invasion; or the Soviet judges whose ignominy put the brakes on biology in their country in the Lysenko affair.

In this loser-takes-all game, then, it is no longer the condemned person who is doomed. Our history is an appeals court in which these condemnations backfire against their judges. Subject to the vagaries of place and power, local laws once carried the day, but the collection of appeals to these judgments created the time we live in: the history of science is driven by the continuous reconsideration of these trials, always in the same direction. This is one of the secrets of Hegel's philosophy: the progressive realization of the reign of spirit, that is, of the sciences, has dialectics, that is, the logic of tribunals, as its temporal law.

At first laws prevail over the sciences, trial after trial; then science prevails over laws, since each is reconsidered in the light of reason; but law prevails, since the internal logic of history, even of the sciences, remains that of the law; but science prevails, since it always delegates experts to the courts; but. . . . The meta-polemic of science and law, of reason and judgment, cannot be decided definitively and constitutes the time of our history.

In the overall balance sheet, traditional history debates endlessly about knowledge and law, about the laws of knowledge of the worldwide world confronted with the laws that organize the worldly world. It is an opposition between two kingdoms, that of this world and that of the other world, whatever the other world may be.

That being the case, one can understand the profound divorce within which we are struggling, with no possible way out. On the one hand, history always vindicates scientific reason, watched over by its founding heroes, all of whom were victims of judicial error and died innocent. The other world, the objective world, thus has reasons that arbitrary reason and the arbitration of the collectivity, being definitively disqualified, have no business knowing. All the battles that scientific tactics lose locally rebound into a global triumph in the war waged by scientific strategy. Yes, science is prevailing over law; and that means that the laws of the world of things are prevailing over the laws of the world of men. In the end that will mean that people will look down on the world of men.

But, on the other hand, this long war is still called history, and its law is dialectics, or the logic of tribunals, which has nothing to do with the world, only with the exquisite disputes indulged in by refined men among themselves. So law prevails over the sciences, even globally, and that means that the laws of the world of men prevail over the laws of the world of things. In the end that means that people will look down on the world of things.

The great legislators of one world are unaware of their counterparts in the other. Must we reconcile two types of law, two legislators; must we bind together two worlds?

Galileo

Eppur, si muove! Condemned, Galileo objects or seems to lodge an appeal: but before what tribunal? Translating his famous exclamation into French makes it apparent that it opposes the affirmation of movement (*elle tourne,* it does move!) with an adverb (*cependant,* yet), which, on balance, designates suspended repose. But there is no established jurisdiction for the new mechanics.

The cardinals decide and pass judgment in the name of canon law, of Roman law, and of Aristotle, the physicist-jurist. To respond to them, Galileo tries to escape from these texts and conventions by positioning himself outside their laws: "my kingdom is not of this world," he says, in substance, or, changing point of reference:

"the world is not within the jurisdiction of this court." He is appealing to a nonexistent authority.

Is the court right or wrong? What does it matter? Since justice speaks performatively and since what it says begins suddenly to exist by the sole fact that it says it, since justice gives rise to jurisprudence in any case, what indeed does it matter, here, to be wrong or right? Judicial truth is its own point of reference; it is founded on itself. If it were otherwise, it would be necessary to ask every court the following question: by what right are you judging?—and thus to form, behind the court, a new authority that . . . and suddenly we are engaged in an infinite trial. No. A given judge pronounces the law by virtue of having the right to pronounce it: this closes the infinite regress into a circle and is called competence.

Galileo's reply, like that of Jesus, calls into question the competence of whoever is judging. And both Jesus and Galileo claim that there exists another space, a supernatural kingdom outside this world, a natural earth in movement, which can serve as a point of reference; thus they challenge the authority before which they are appearing and whose competence concerns criminal or political affairs in the first case, and canonical ones in the second. It is always true that for every law there exist spaces of non-law, where the conventions are different: there this court has no competence. This term is used, even in the sciences, for the right to judge, that is, the right to exercise jurisdiction. So appeals are directed to another competent authority.

If the court in session can justify the legal basis supporting its judgment, the objector cannot, since the legal basis of his position, by definition, does not exist, at least not yet. If it existed, the defendant would not of course refer to a space of non-law; his case would come up as a routine matter. The sitting jury is thus justified in demanding from him some sign or testimony that would make this space of non-law seem plausible or would at least point to it.

In response, the hero can choose to display or not the existence of things outside the text, things on which he bases his challenge to the text of the law. Conversely, the court demands that he account for these things that go beyond the case before it: the space of the law includes things that are equivalent to cases or cases that are the same as things, whereas the space of non-law contains things that are not cases, that are not yet cases, or even that never will be. This reservoir of references, of things to refer to outside the law, can be called transcendence. In Roman law, the defendant is called *reus* and his cause or case *res.* Let us call *real* the space to which he refers, which cannot be justified by any text.

Finally and above all, Galileo, like any other author, needs some tribunal so that his theory, which is probable, becomes canonized, so that his real becomes rational and his text attains the status of truth. Unable to decide between equivalent astronomical hypotheses, and in the absence of an *experimentum crucis,* a decisive experiment, celestial mechanics requires a final judgment in the court of last resort. Science always requires that.

Eppur, si muove! Galileo objects, lodges an appeal. He invokes the world of things themselves, the earth and its rotation, peaceful, uncaused, no court case. Religious magistrates preside over the Husserlian Earth that does not move, and over the Heideggerian Earth that envelops them and gives them a foundation. Confronting the cardinal, the astronomer carves out two spaces, of law and non-law: from the first, contractual or conventional, appeals are made to the second, natural. In this contractual space, an Earth that moves seems as alien as a global change does today: is it a thing without or with a cause, a case that can't or can be made?

Filing an objection on the question of competence, the physicist lodges an appeal to this nature. To a natural law. Which is going to be born, but which is not yet born. There are no performatives in nature.

Is it really a space of non-law, or simply a court of appeals? Jesus appeals to another world; Galileo, too; but so do Hobbes, Montesquieu, Locke, and Rousseau. Jesus and Galileo questioned or weighed a given precise judgment pronounced by some particular jurisdiction, but since then legal philosophers have pondered the general issue of man-made laws, contracts, or conventions, in order to found or engender or amend or destroy them. Like Jesus and Galileo, they appeal to nature, which is required to decide and judge, as a court of last resort. Without convention, self-foundingly, transcendentally. Here, without contradiction, are mingled the natural and the supernatural; factual history and formal, general, or logical conditions; deists and atheists—all hearkening to the ultimate reference point, beyond which no further appeal is possible.

This court of appeals or of last resort pronounces such weak, general, and contradictory judgments, in the work of respected philosophers, who are perpetually suspected by their contemporaries and successors of confusing this court with their own conventions, that modernity closes down natural law and no longer has at its disposal a back-world or other point of reference. Instead it can only have recourse to the fluctuating decisions or the gaudy violence of historical circumstances. Obligated, by the obliteration of any other world, to do without the possibility of appeal, we collapse or contract together the first and last authority, and this contraction defines us all. We survive among man-made laws, tossed about by the history of dominations.

During this time, Galileo's appeal has remained on file, but it has encountered no competent court within what is still called law. And the nature he summons becomes that of mechanics and its competence. Then natural law is identified with the physical sciences: they take the place it leaves. We no longer refer to anything but knowledge's expert testimony. Thus it is that we know, but can no longer decide.

Science alone now has all the rights. Originally, law preceded it; during history, these two authorities opposed one another, each assuming the other's prerogatives in turn; in the end, science, which alone is competent, is left holding the field, and the Earth.

We thus forbid ourselves new messiahs or Galileos. Unless we reopen this closed-down nature, unless we invent a new global natural law. For the Earth is now holding us.

Because of its dazzling success, science now occupies the space of natural law. Galileo appeals to the Earth that moves, whose movement could not, in the eyes of the jurists of his time, assure a fixed reference for any judgment; this appeal gives something like an ownership contract to exact knowledge's conquest of the terraqueous globe.

Galileo is the first to put a fence around the terrain of nature, take it into his head to say, "this belongs to science," and find people simple enough to believe that this is of no consequence for man-made laws and civil societies, closed in on human relations as they are. He founds scientific society by giving it its property rights: in so doing, he lays the deep foundations of modern society. The knowledge contract becomes identified with a new social contract. Nature then becomes global space, empty of men, from which society withdraws. There the scientist judges and legislates, mastering this space where man-made law had more or less left technicians and industrialists alone, free to go about innocently applying the laws of science—until the day when the natural stakes began to weigh more and more heavily on man-made debates.

Nature lies outside the collectivity, which is why the state of nature remains incomprehensible to the language invented in and by society—or that invents social man. Science enacts laws without subject in this world without men: its laws are different from legal laws.

The experimental sciences make themselves masters of this empty, desert, savage space. Philosophers thought that such space, if it existed, contained the conditions of possibility, the source, the foundation, the history, the genesis, the genealogy of all law, and even implied its distribution among several authorities, all of whom respond to the indefinite question "by what right?" and converge toward a final reference point. By becoming owners of the space of non-law, the sciences, with their competence, furnish experts for courts and thus decide before they do, and in their place.

Natural law is dying because science has conquered its space. Science plays the role, now, of our Last Judgment. Henceforth law and science are opposed as the man-made and the natural once were, always to the benefit of the natural. One result of Galileo's trial is that reason without a subject, objective reason, prevails over the reason a subject can speak, so it decides without you or me having anything to do or say.

How could we not recognize again, in Galileo's trial, the immemorial biblical debate of the prophets and the kings? Those who live by law demand of the new arrival, who claims to speak of another world, a miraculous sign showing truly that he comes from elsewhere, from God or another world.

Then, by raising his hand, the physicist sets the whole Earth in motion. Subpoenaed or cited by the court, he cites the Earth, appeals to it, and makes it move—as we know, the verb *to cite* means, in ancient languages, to shake. An immense astonishment that will change history: it is moving! What is a miracle? The thing suddenly bursting in on the case, the world barging into the courtroom: an earthquake. De facto, it is shaking! This is the original, truly miraculous status of mechanics, the new science of movement. The phenomenological Earth was shaking.

We have not gotten over it yet. The prophet overthrew the king. Science takes the place of law and establishes its tribunals, whose judgments, henceforth, will make those of other authorities seem arbitrary. And now, what are we to do and how are we to decide, by what right, in a world and a time that only knows how to know and only does that which ensues from knowledge? a world where science alone is believed? where only its courts judge in a doubly competent way, uniting law and non-law?

But now we encounter something new. At the borders of effective and precise knowledge, and at the limits of rational intervention, we find not only ignorance or error but mortal danger. Knowing is no longer enough.

For, as of today, the Earth is quaking anew: not because it shifts and moves in its restless, wise orbit, not because it is changing, from its deep plates to its envelope of air, but because it is being transformed by our doing. Nature acted as a reference point for ancient law and for modern science because it had no subject: objectivity in the legal sense, as in the scientific sense, emanated from a space without man, which did not depend on us and on which we depended de jure and de facto. Yet henceforth it depends so much on us that it is shaking and that we too are worried by this deviation from expected equilibria. We are disturbing the Earth and making it quake! Now it has a subject once again.

Science won all the rights three centuries ago now, by appealing to the Earth, which responded by moving. So the prophet became king. In our turn, we are appealing to an absent authority, when we cry, like Galileo, but before the court of his successors, former prophets turned kings: "the Earth is moved." The immemorial, fixed Earth, which provided the conditions and foundations of our lives, is moving, the fundamental Earth is trembling.

This crisis of foundations is not an intellectual crisis; it does not affect our ideas or language or logic or geometry, but time and weather and our survival.

For the first time in three hundred years, science is speaking to law and reason to judgment.

Historical Meetings of Science and Law

Yet they have sometimes lived together.

Examples. Aristotle defines justice by the law of equilibrium, whose technical model is expressed by the figure of the scales, and whose universal equation is given by the proportional analogy a/b = c/d. Were there, in the ancient Greek world, two more general statements than the design of the most complex of the simple machines and the most efficient "algebraic" method? Distributive justice has already left behind strict equality, which is too naive, and turns to compensation: two unequal weights are balanced through the unequal length of the arms of the scale; many differences are thereby respected. The highest science of the time expresses the best law.

More than two thousand years later, Leibniz invents the integration of differential multiplicities. Of course there are differences, but integral calculus provides for them. The most global sum is always the most correct, because it leaves intact the most pluralities. This is the most general method of the age, and here is its technique: everything in nature follows the extremal paths defined by the calculus of variations. God creates mechanically the best of all possible worlds, just as falling bodies follow the greatest slope and the sphere of raindrops conforms to the greatest volume or the pendulum conforms to the curve of the least time. This is a decisive argument for saving God, before the tribunal of men, from the accusation of having created evil. The most general natural law is applied to the universal juridical problem and solves it.

Judging is equivalent to weighing: as acts, they have calculation in common; as words, they share the meaning of *thinking*. Aristotle's justice seeks a certain compensated mean, whereas that of Leibniz's God follows the extreme edges. The two theories regulate the universe through the singularities of mean and extrema.

These laws of nature almost always come down to expressions of equilibria or of invariance through variations, to structural laws, including those that give the largest share to time, the laws of evolution. We could call them, literally, laws of justice. In these cases, the fluctuating and differentiated equilibria of inanimate multiplicities and of variable but defined species are equivalent to equity in collective situations.

Natural accuracy, or justness, then, raises exactly the same questions as does social, legal, or moral justice. This natural law, inspired by the natural sciences, and whose broad outlines are shared by a now global technology, this natural law is not unlike human laws but remains parallel to them.

As the history of science advances, the notion of equilibrium progresses and becomes refined, capable of integrating more and more disequilibria into a broader and broader concept. Invariance without differences is foolish: Plato makes us laugh by his inability to conceive that a top moves less on its base the faster it turns on its axis; the thing seems contradictory to him. By the proportional analogy, on the other hand, Aristotle subsumes the inequality of the scale's arms under a principle of strict equality. From Aristotle to Leibniz, we pass from statics to the calculus of variations, in which stability takes into account a certain movement. This tendency will continue forever: a former immobility assimilates the most turbulent variations, as if a race were developing between an ever broadening statics and the complete set of conceivable movements. A new deviation shakes up a system, which some new invariance then restores to calm.

A chreode, for example, brings to the fore the global equilibrium of a mobile flow: if you move a river's normal bed sideways, it will come back toward its former channels; the orbit of movement itself seeks repose. As for chaos theory, it distributes its attractors along fractal curves, whence the discovery of a refined order beneath the appearance of the most disturbing disorder. This could be a good theory of history. Here then are broader and broader concepts that make us understand constancy in movement, or, beneath the apparently chaotic paving-stones, the sandy beach of a distribution.

I imagine that climate refers, in the same way, to general invariances that absorb the brief devastation of the most sudden hurricanes and the slowest cycles of marine currents. We do not yet know all that is covered by the term *global change,* nor even whether this designation has any meaning. One can imagine that the most drastic changes might end up being integrated into a fairly stable higher synthesis, one that brings together physical questions and social or political problems: then, although both are chaotic in the most refined sense, geography would comprehend history and history geography. Can one think, estimate, and calculate in order finally to steer the changes of Planet Earth, without integrating into a global model all the local modelings and their constituent parts, combining variables both natural and human? It is always the same question, of invariances and variations, of disorder and order, brought to the highest level of integration. Like philosophy long ago, science is finally thinking universally, both keeping and losing all the divisions that historically created its power and efficacy, because it is trying to bring them together. Science thinks by casting off from the local to the global.

The idea of justice designates precisely the ideal limit pursued by a continuous work of expansion, through which an equilibrium is able to absorb ever larger deviations, while allowing them to subsist. So it could be said that the history of science follows, in this respect, the series of legal appeals from the local toward a global.

In short, could all of science be practiced without one or several general constants that assure the regulated functioning of reason? As if these constants referred at last resort to the fundamental, immutable Earth, which scientific work distributes into multiplicities of variables expressing properties or positive laws?

Does science have the same foundation and the same aspect as law? Is there thus a single reason, which would be distributed into domains attributable respectively to justness and justice?

Principle of Reason

Leibniz states in its Latin form the *principium reddendae rationis,* according to which not only does each thing have its sufficient reason, but also reasons must be given back or *rendered.* As we know, this principle founds scientific knowledge and thus justifies its name.

To my knowledge nobody has noticed the use, here, of the verb *render* from the pen of one of the eminent jurists of the era. This rendering or return expresses either a reciprocity or a consequence with respect to a prior action and thus implies that he who renders must first have received some gift. The principle of reason requires him to do it; it thus establishes the usual equilibrium as far as contracts are concerned, and is founded on equity in exchanges. It's an equation of optimization, symmetry, and justice, and so there must be, preceding it, a real or virtual contract. Reason is founded on a judgment.

But who gives what, and to whom must we render reasons? The answer leaves no doubt: to all things. If every thing has its sufficient reason, we must render that reason to the very thing, well named, that we call the given. The world, globally, and phenomena, proximate, local, or remote, are given to us; it would be an injustice, a disequilibrium, for us to receive this given free, without ever rendering anything in return. Equity therefore demands that we render at least as much as we receive, in other words, that we do so sufficiently.

What can we render to the world that gives us the given, the totality of the gift? What can we render to the nature that gives us birth and life? The balanced answer would be: the totality of our essence, reason itself. If I dare say so, nature gives to us in kind, and we render to her in cash, in human sign currency. The given is hard; reciprocity, soft.

The principle of reason thus consists in the establishment of a fair contract, the one we have always agreed on, the one we observe in real time with nature.

The principle of reason describes the natural contract, which embodies both reason and judgment.

At the time of the classical rationalists, the principle furthered only the concern to establish laws: those of physics or other natural sciences are subordinate to the principle of reason, as are the laws of any particular man-made legal order with respect to the universal and quasi-natural principle of the equity of exchanges or the equilibrium of contracts. Thus positivism or even rationalism are philosophies with a juridical foundation.

This rational contract, which balances the given with reason, brings to a close the transhistorical conflict between the world and us, a war marked by a thousand defeats for a few rare victories, and as many strategies of sham obedience and real command.

Thus this contract expresses a pact, a sort of armistice; we come back to the war with which we began. We would never have signed it as long as we were being defeated during these confrontations. Before the contract, the given gave us more harm than gifts, and we found ourselves in thrall to nature. Thus the contract inaugurates a new era, during which we will submit the world to a reasoned inspection. Of course we render reason to it first, but we are also bringing it forcibly to reason. Rationalism and positivism are crowing over their victory. The soft is getting the better of the various hardnesses. The world is entering the book. The armistice closes a war won by reason.

To the verb *render,* which comes from law, is added the word *reason,* which also comes from law, because it signifies proportion, distribution, moderation in equilibrium. The contract established by the principle of sufficient reason would not be totally rational if it were not reasonable as well. One must surely render to nature neither less reason than the given demands, nor more. If reason exceeds the given, the contract is broken, as surely as for the opposite reason. The principle requires that an equilibrium be reached. In the same way, a necessary condition becomes sufficient, as well, if and only if the implication that joins it to the conditioned term turns back, in balanced reciprocity, from the conditioned toward its condition. In a way this double arrow displays an equilibrium.

In Leibniz's time the principle of reason was the expression of the rational contract founding the natural sciences, as if reason were finally reaching the point of balancing the given, after a long period in which reason got the worst of it. Now, conversely, the given itself is disappearing under the weight and power of reason's productions. Thus we tend to reinterpret the principle of reason as a reasonable contract.

Why call it a natural contract? In Leibniz's day, the lawyer in this case pleaded on the side of reason and never in favor of the given, which was so profuse that it inundated us from all sides. In a way, nature itself forced us to render reasons, as one makes a loser surrender all he has. Today, we ourselves, reasonable men, are brought to plead on the side of the given, which, for some time, has been laying down its arms. The book is going back into the world without the world leaving the book.

The principle of reason comes down to a rational contract when reason tips the balance in its case against nature, and would conversely come down to a natural contract, if nature, through our voices, tipped the balance as much in the case that opposes it to reason. Through reasonable reason, the principle of reason brings balance to its reason. Through moderation, it distributes power equitably, since reason means at once the excess of power and its limitation. In this principle, finally, the rational sciences return to just law, and reason meets judgment.

Acting as the Good Lord's lawyer in the case brought by men against Him, on the problem of evil, Leibniz completed his *Theodicy*. Similarly, as a defender of reason and friend of the truth given by God, he began, with the principle of reason, this *Epistemodicy* that we have continued, explaining the relationship of reason to judgment, so inevitable that God himself could not escape it.

The question of evil is being raised once again, given the responsibility of our sciences, our technologies, our truth. What are we to do about it?

Some philosophers, including Leibniz, are lawyers by vocation; others are prosecutors, like Socrates and our contemporaries in the social sciences, who are ready and willing for police duty; others, finally, judge, like Kant . . . In Greek, Paraclete, the lawyer, carries the name of the Holy Ghost; and in Hebrew, the prosecutor is called Satan.

Can philosophy escape this courtroom? What should be said there today, when science is speaking to law and reason to judgment?

Reason and Judgment

Let us distinguish two types of reason, or, in other words, distinguish reason from judgment. For the first, which presides over knowledge and then science, the necessity of truth comes from fidelity to the real or from proof. Here truth overturns error, the misinterpretation or shadows borne by the imaginary. Since the Enlightenment, this reason has lit our way, in theory. Deprived of it we would think falsely. For the second type, which presides over legal reason, the need for arbitrating or, worse yet, for the arbitrary, comes from violence and death. Without the arbiter, we would be exposed to the worst risks, we would kill one another. Justice is competent to know cases and causes, and justness or accuracy is competent to know things.

From error ensues truthful reason and from death ensues judgment. To defend ourselves temporarily from death, and to claim to leave error behind once and for all, we need both reasons, faithful knowledge and prudent judgment.

Since the risk of error ultimately made us run lesser perils than the danger of death, we rightly used to put judgment above reason and law above science. Judgment characterizes tradition and reason characterizes innovation. From long experience, the old man loves prudence, whereas the young man reasons.

The rising power of the exact sciences overthrew this state of affairs because their efficacy began to preserve us from death, through technology and medicine. Starting with the Enlightenment, reason has presided at the court of judgment; expertise tilts verdicts decisively; the great scientist garners the glory that formerly brought fame to the legislator; rational or experimental youth prevails over reasonable and experienced age. Above judgment rises reason.

Now we are witnessing judgment catching up with reason. The successive crises of the sciences and their affiliated technologies, each of which, at the apex of its power, came close to mortal danger—atom and bomb, chemistry and environment, genetics and bioethics—these crises bring back the demand for prudence, as the helmsman of what is effective and what is true. We were once old, more recently we've been young, now we are mature. Why should human history follow the same course as organic life?

Today, our collectivity can equally well die of the productions of reason or safeguard itself through them. Reason, which used to decide, can no longer make decisions about itself. It appeals to law. And our judgment cannot do without the productions of reason. It appeals to the sciences. Our philosophies cross, and this is the cross they have to bear.

There is no contradiction here, but a positive cycle. Thus it is better to make peace by a new contract between the sciences, which deal relevantly with the things of the world and their relations, and judgment, which decides on men and their relations. It is better to make peace between the two types of reason in conflict today, because their fates are henceforth crossed and blended, and because our own fate depends on their alliance. Through a new call to globality, we need to invent a reason that is both rational and steady, one that thinks truthfully while judging prudently.

Yet we no longer believe in faculties of conscience, reason, and judgment, sitting next to imagination and memory among other similar functions or organs in a bright-dark vessel, nor do we believe in concepts equipped with lofty capital letters; but we know real people, and real people are what have to be invented. To form them, we need an education, and for that, a model. Let us then trace a portrait that never had a precedent, in order that it may inspire imitators.

The Instructed Third (*Le Tiers-Instruit*)

Today's Sage is a mixture of the Legislator of heroic times and the modern titleholder of rigorous knowledge; he knows how to weave together the truth of the sciences with the peace of judgment; he blends together our Egyptian and Roman heritages, the source of our laws, with our Semitic and Greek legacies, givers of knowledge; he integrates quick and effective sciences into our slow and prudent laws. Young and old at the same time, the Sage is reaching maturity.

I call this Sage "le Tiers-Instruit," the Instructed Third, knowledge's troubadour: expert in formal or experimental knowledge, well-versed in the natural sciences of the inanimate and the living; at safe remove from the social sciences, with their critical rather than organic truths and their banal, commonplace information; preferring actions to relations, direct human experience to surveys and documents, traveler in nature and society; lover of rivers, sands, winds, seas, and mountains; walker over the whole Earth; fascinated by different gestures as by diverse landscapes; solitary navigator of the Northwest Passage, those waters where scientific knowledge communicates, in rare and delicate ways, with the humanities; conversely versed in ancient languages, mythical traditions, and religions; free spirit and damned good fellow; sinking his roots into the deepest cultural compost, down to the tectonic plates buried furthest in the dark memory of flesh and verb; and thus archaic and contemporary, traditional and futuristic, humanist and scientist, fast and slow, green and seasoned, audacious and prudent; further removed from power than any possible legislator, and closer to the multitude's ignorance than any imaginable scientist; great, perhaps, but of the common people; empirical but exact, fine as silk, coarse as canvas; ceaselessly wandering across the span that separates hunger from surfeit, misery from wealth, shadow from light, mastery from servitude, home from abroad; knowing and valuing ignorance as much as the sciences, old wives' tales more than concepts, laws as well as non-law; monk and vagrant, alone and vagabonding, wandering but stable; finally, above all, burning with love for the Earth and humanity.

This mixture demands a paradoxical rootedness in the global: not in a plot of earth, but on Earth, not in the group, but everywhere; the plant image hardly makes sense anymore. Since we left the ground, casting off powerfully for remote places, we have relied more on immaterial bonds than on roots. Could this then be the end of all forms of belonging?

Rearing

May this Sage found a lineage. The rearing of the human baby is based on two principles: the first, positive, concerns his instruction; the other, negative, involves education. The latter forms prudent judgment and the former valiant reason.

We must learn our finitude: reach the limits of a non-infinite being. Necessarily we will have to suffer, from illnesses, unforeseeable accidents or lacks; we must set a term to our desires, ambitions, wills, freedoms. We must prepare our solitude, in the face of great decisions, responsibilities, growing numbers of other people; in the face of the world, the fragility of things and of loved ones to protect, in the face of happiness, unhappiness, death.

To deny this finitude, starting in childhood, is to nurture unhappy people and foster their resentment of inevitable adversity.

We must learn, at the same time, our true infinity. Nothing, or almost nothing, resists training. The body can do more than we believe, intelligence adapts to everything. To awaken the unquenchable thirst for learning, in order to live as much as possible of the total human experience and of the beauties of the world, and to persevere, sometimes, through invention: this is the meaning of equipping someone to cast off.

These two principles laugh at the paths that guide today's contrary educational practices: the narrow finitude of an instruction that produces obedient specialists or ignoramuses full of arrogance; the infinity of desire, drugging tiny soft larvae to death.

Education forms and strengthens a prudent being who judges himself finite; instruction by true reason launches this being into an infinite becoming.

Earth, the foundation, is limited; yet the casting off from it knows no end.

Casting Off

The Port of Brest

Eve is a blond sporting a short, black-and-white rose print dress; her acid green shoes match her belt. Dark-haired Adam is shivering in navy blue pants and a jacquard sweater. They have a good kiss. The cold October wind whistles, pinning the boat to the quay. Everyone's waiting for it to cast off.

Once the gangplank is in place, very steep because of the high tide, the passengers get on board. They have a hard time of it, dragging their bundles and their children. Unsteadily, they hand their tickets to a sailor, who gives each of them in turn a kind and cheerful look. It takes time for everybody to get set, some down below to avoid the cold, some on the foredeck to feel the sea breeze.

Once the gangplank is pulled away and the guardrail latched up, the fore spring and the bow breast line are loosed. The sun is just rising. Eve has stayed on land, and she turns laughingly toward her boyfriend, standing in the center of the deck; Adam looks down at her, his feet on a level with her head. From her purse she takes a big red apple and bites into it. The prow is already starting to angle away from the pier. Cradling his hands together, Adam gestures to Eve to throw him the apple. She tosses it, he catches it. She bursts out laughing again.

Bound for the Ponant Islands, the *Enez Eussa* is slowly pulling away from the quay. The stern hawsers are still in place. Adam munches on the fruit and tosses it back to Eve's hands with a smile. Now the sun is up and the passengers can start noticing things besides their own queasiness. Smoke from the stack sweeps over the deck and then drifts off in the set of the wind. The girl catches the big red bitten apple, looks at it, hesitates a moment, and boldly sinks her incisors into it. The stern breast line drops into the water, the sailors hauling it in and stowing it. The rear of the boat pulls away as the apple flies through the air a third time, from her to him. The engines pick up the pace; the boat heads for the mouth of the harbor. The apple, getting smaller, passes from sea to land once again.

Adam and Eve are no longer laughing; far from it, they're in a hurry. Toss, wait, catch, bite, throw back. Seated astern, I watch the couple's little game, which started out just like that but has become hurried, urgent, difficult, and I'm losing count. Tracing ever-longer arcs as it gets smaller, and as the boat, getting up steam, heads away sounding its horn, the apple soars ever more majestically. Seriously, even assiduously, the two lovers pursue their absorbing task, concentrating on it so intently that they don't notice they've become a spectacle for the sailors and some of the passengers. From quay to shipboard, from deck to pier, the apple keeps stubbornly spinning along, like a living arrow, spinning ever larger and looser bonds between the hands separated by departure.

I could swear that because of the spiderweb spun by the fruit, going and coming like a shuttle, the ship is having trouble getting away, still tied to land by the visible and invisible hawsers of wafting memories and regrets. There's nothing stronger, you know, than these araneidan threads. Through how many heartrending goings and comings does the messenger rise and spin, ever lighter with each successive flight path?

But the craft has finally cast off, the fruit is consumed, and when the core is down to the seeds, the broad parabolic arc, which ought to drop it into waiting hands ready to throw it back, instead misses and sends the apple right into the dirty water.

Without a sign, Adam and Eve turn away, a pair no longer. At that distance, no one can recognize anyone else's body.

Circling gulls dive down to fight over what remains afloat of all that good will. Seeds.

Kourou Base

Delayed twenty-four hours by an imaginary breakdown falsely detected by the computers, the blast-off has just been ordered: three, two, one, zero. Is the spacecraft *Ariane* casting off?

The clouds and the glow appear first. When the sound hits, the ears can't believe it. No, this isn't the noise of any known motor: between the groaning mangroves and the edge of the forest, deep in the equatorial night, comes an event on the scale of weather, not of some man-made technology; a storm, a typhoon, a hurricane, a cyclone is passing over us, what our ancestors called, precisely, a meteor: God's thunder, the lightning bolt, gusts and clouds.

For a few passing moments we lose sight and sound of this scourge of the atmosphere. The lightning flame is now becoming a signal, then a bright point, taking its place amid the thicket of stars. Second-stage ignition: a comet appears for a moment. We try desperately to follow it in the night sky. Within a minute the new planet lights up. As we watch, *Ariane* now belongs to astronomy. Casting off, its launch has just linked the lower regions of the air, where disorderly meteors seem to reign, with the heights of heaven ruled by the order of the stars.

When seafaring craft cast off, they turn their antennae toward a world that is strange beside the landlocked daily routine: on the plain of the high seas, nothing ever resembles what's been left behind. What's square becomes round, what's stable moves; you'll never make the same movements, you'll speak a singular language, which no one who hasn't been there will understand. To leave is to sever all bonds.

To go out from this world and enter another, where nothing will be the same: that's called casting off. Equipped with their gear, foreign to land and adapted to the sea, loosing their hawsers and cutting the fabric of former connections, vessels are capable of providing this shattering transition. We're going to live differently, perhaps for a long while, elsewhere, where the watchman will have only the wind and sky for companions; that's why sailors always have about them, when they return, that odd little air.

What clairvoyant and melancholy genius, then, composed the bugle call for casting off? It's more heartrending than taps at a funeral.

If a car passes through town, it's doubtless on its way from Toulouse to Bordeaux; Paris and Madrid are connected by this airplane humming over our heads. All this sound and vapor, this rude and filthy noise for a merely superficial change. Lolling in a vehicle from which they never look out at anything, and which never leaves the obligatory corridors and checkpoints, the passengers read the paper, anxious not to leave their space or their time or the ordinary murders with which the news drugs them.

Here at the edge of a forest that's hard to get out of alive, at the edge of this other world called primitive, hurricane *Ariane* is taking a communications satellite into space, and in the process linking the chaotic sky of meteors to that of astronomers, the ordered space of celestial mechanics.

Now, if Bordeaux and Madrid depend somewhat on us, since our ancestors founded them and we fancy that we run them, neither climate nor constellations ever did. Neither sky nor seasons are of our making or unmaking.

Antiquated, low-technology vehicles go from this world to more of this world, from a city to a capital, without ever leaving the guidance of roads, and even roads have recently become mere streets, since the monotonous and dominant model of the city is ruthlessly overrunning space. From Milan to Dublin, megalopolis Europe reigns.

As for boats, they cross over from this earthly or landlocked world to the other world of the sea. And *Ariane* passes from a world of otherness to another world, a still more difficult passage. Right from its departure, it's in the thick of instability and chaos, the uncontrollable space of storm, thunder, and lightning; it unleashes the most volatile elements, fire and air, in the lower regions of the atmosphere, so as to reach the ordered heights that have always escaped our grasp and our enterprises.

Kourou, 1 April 1989, 11:29 P.M. I look back at the few spectators, clients invited to the launch: tears glisten in everyone's eyes, as I discreetly hide my own. Engineers, scientists, and experts present themselves, as far as I know, as cold and reasonable men, having long since become blasé through repeated calculations and projects. Yet they are crying. I suddenly thought I was seeing them come naked out of the forest to be dazzled or terrified by this comet and cyclone, like wild men who know perfectly well that we can do nothing about stars or hurricanes.

Before our eyes and ringing in our ears, the lightning and thunder of a storm have become a new planet. Suddenly we've changed back into what we had never ceased to be: primitives. By the vigor of its thrust, the highly sophisticated achievement reawakens what is archaic within us.

We remain plunged in our distant past, without noticing it, up to our thighs, our shoulders, our eyeballs. Panic-stricken, we were watching an ancient ceremony whose splendors celebrated both the calm constellations and the natural forces that burst and blast; we had cast off for a forgotten era of our prehistory, going backward in space and time. An action rising up toward the future means a reaction downstream, a shakeup in the foundations.

This feat subjects us to the longest and darkest of recollections: yes, we are archaic in three-fourths of our acts and thoughts. Launched toward the farthest places, we find ourselves flung into primitivity, as if our casting off, which unties one set of bonds, had retied another. The process of hominization "takes" in us, the way a crystal undergoes a phase change and solidifies: does becoming human consist of forever unbinding so as to bind elsewhere and otherwise? Do we cast off only to change cords?

The looming forest—another world, and doubtless our wild origin—touches us, surrounds us, permeates us, and doesn't leave us. Perhaps we never stop going home to this third world (which is still first) so that we may leave it, or saving ourselves from it so that we may go back to it. The most advanced of men are rooted in the farthest and darkest traditions.

Chabournéou en Valgaudemar

Three in the morning. Silently, everyone gets up, folds the sleeping bags, eats a quick breakfast, and leaves. Attentive, courteous, and diligent, the hut-keeper hands out canteens of weak tea, surveys the roped parties forming up, and checks off the destinations. Outside, the darkness is pricked by little dancing fireflies, the headlamps, a dawn before dawn. In the giddiness of night, each takes refuge in his personal glimmer and his four inches of path. All solitaries.

Before the nighttime vigil in the shelter, none has left this world; with the wee hours, each is entering the other world beyond. This little hut near the glacier serves as a checkpoint, a door, an airlock, an entry, a passage, guarded by a kind of Saint Peter. Snow, ice, and rocks make up the other world, a near abstraction. It has nothing in common with the usual world. Horizontal becomes vertical, our old forms of stability stir, all gestures and behaviors change. Language is transformed, becoming something that won't be understood by anyone who hasn't been there. You can walk sixteen hours for the extraordinary reward of facing the wind and sky on a summit among the summits, which are like raised arms, observation bridges or standing trees. On returning from the trek, easy or demanding, everyone's eyes conceal a strange, wild air: a ruddy gleam that's the signature of the place's uncanniness.

Ever since our first parents were expelled from paradise's garden, we must all be bearing a mark like this that we no longer notice.

The archaic and the primitive accompany us step by step. Have I mentioned yet that glaciers break into crevasses when their beds become convex and slope downward? Everyone knows that. White, wan, and green, the lips of these mouths, or bergschrunds, gape, here and there bridged with snow.

When these same beds rise into a concavity, as sometimes happens, the glaciers break across their thickness, but in the opposite direction, so that the crevasses take on the form of an inverted *V*. You cross a barely visible line, squeezed tight, solid, and locked under enormous pressure, but its narrowness conceals a gigantic volume that grows as it gets deeper, and in some cases could contain several cathedrals.

High and white, the mountain hides giant spaces, low and dark. There, it's said, sound faints away, appeals for help are lost, light dimmed. No lamp can light them; no one has ever come back. Unseen, unspoken, some lowly past shadows and prods every upward voyage.

The high seas and high mountains are like the high heavens in that you must cast off to reach them: you have to go through the port, the mountain hut, or the launchpad. These vertical chimneys lead upward via a bizarre labyrinth, where the guide, as in the era of the murderous Minotaur, is still named Ariadne, *Ariane.* In all these often interminable voyages, every passage resembles the mazes of ice near the Northwest Territories.

But if elsewhere every departure presupposes that threads and bonds be broken or that hawsers be undone, the wee-hours departure from high-altitude shelters requires, for its part, the formation of roped or corded parties. Hardly anyone ventures up there alone. A continuous but supple material communication forms between the climbing harnesses, assuring forward movement. The subject that walks, scales, ascends with crampons, gets through or doesn't—this subject is not you or I or he, but the roped party, that is, the cord. You may be an anchorite who has emigrated to the remotest dens of high silent valleys, but now you have cast off, doubtless in spite of yourself, for something collective. The subject who will be dazzled by the mauves of dawn amid sheer corridors will be the love that your guide and your friend prove to you in their every movement and step, and that you in turn bear them: in other words, the cord, once again. Call it cordiality, the apple of concord.

The term *contract* originally means the tract or trait or draft that tightens and pulls: a set of cords assures, without language, the subtle system of constraints and freedoms through which each linked element receives information about every other and about the system, and draws security from all.

Thus the social contract itself, in the form of a cord, moves about or scales sheer ravines, from dawn to noon: you'd think it was any collectivity going by, bound by the obligations of its own laws and by the world.

Take another look at the vessel casting off: in slipping its hawsers it comes loose from but a small part of the web, the network, the complex tangle of bonds that hold it and that have a name only in sailors' talk. Unbound? No: bound fast. Do we leave one contract only in order to contract others? What was a sailing ship if not an exquisitely complicated gigantic knot? The group that launches an attack on the rock face is called a roped party: here are two contracts departing for history.

It was once thought that the word "society" derived from the French verb *suivre,* Latin *sequere,* to follow, and thus took on the form of a sequence. You're a hermit or libertarian and you dreamed of leaving every kind of collectivity? As soon as you wake up you're back to its pure sequential model in its bare form: sex, the hyphen or trait or draft that unites, the umbilical cord.

To protect themselves from danger, alone and closed, like certain crustaceans, the warriors of the Middle Ages and antiquity wrapped themselves in back-breaking armor. Like war, nature later came to prefer the flexible strategy of soft flesh outside and hard skeleton inside. A third solution, more advanced in a whole new way, lies in placing one's defenses and security outside the body: in relations. What comes out of me or hangs from me or leaks from me is what saves me: I'm casting off toward the rope, the cord. Although we have no proof, this bond must have constituted the first invention of human technology, at the same time as the first contract.

Now in soft surroundings, as long as the terrain stays flat, no one feels the need of bonds and each person strolls comfortably about alone, but when the terrain rises up and becomes hard, the collectivity becomes a roped party and takes refuge in the social contract.

If the mountain, finally, turns out to be difficult, appallingly tough, then the contract itself takes on a different function: it no longer binds just the mountaineers among themselves, but in addition anchors itself to the rock face at specific strong points. The group finds itself bound and submitted not only to itself but to the objective world. The piton is an appeal to the strength of the cliff, which must be tested before any bond can be made to depend on it. A natural contract joins the social contract.

These craft casting off—vessels, spacecraft, roped parties—represent the sum of relations that groups must have with the world, when the world, for those who have cast off, is a dangerous one. What relations do these craft, in turn, maintain with law?

Did I tell what the guide told us during one of our few rest stops? One morning, he and another climber, taking turns going a rope-length ahead, were clinging to a north face, three-fourths of the way up a vertical ravine of sheer ice, when, at a moment when they had come even with each other, they heard the telltale hiss of layers of air driven down by falling rock. Experienced mountaineers have sharp enough hearing to detect even the coming of the Holy Ghost.

Side by side across the vertical axis of the icy face, both hands clinging to their ice axes, their crampons planted squarely in the hard wall, instinctively fleeing sidewards, since falling rock usually follows the main path in the center, just where they happen to be, one of them pulls to the right and the other to the left. The cord tautens in proportion to their strength. Both said later that they had felt hatred vibrating along that quivering violin catgut. In seeking cover, each hauls on the other as if to put him in danger. But no: a block as big as a boat comes crashing down between them like a thunderbolt, and in a single blow rips away pitons, carabiners, ropes, and corkscrew pitons, the whole carefully woven mountaincraft in which they had cast off. After the storm, there they still are, alone and safe, stuck to the wall, two flies. The supreme force of their apparent hate had saved both of them by driving them apart.

Separation is sometimes a loving solution.

But adversity always attacks the principal means of defense, namely communication. The rock undoes the roped party, storms rip away the tangle of ties and knots, this network that makes up a boat, the exquisite technologies for casting off, appropriately called craft, leaving them in distress.

Crises tear contracts.

Cord and Bond

Law, the refined technology of our relations, can sometimes be recognized and read in certain expressions that apparently refer to concrete and technological origins. The terms contract, obligation, and alliance, for example, speak to us etymologically of ligatures, ties, bonds: there, our liaisons or connections are once again threads.

A cord, which, when knotted, can hold things fast, strikes me as the first tool, whether it is used for men, animals, or things. Without it, how to bind the stone to the handle or tether the beast to its post, how to tie the prisoners' wrists, weave a loincloth or take to the sea? Or entwine one's lover? The cord is for pulling, drawing. Drawing, gripping: together they make an arm, which can reach, and a hand, which can grasp. The bond remains effective, without the bodily organ, and works on its own.

By its flexibility, which leaves a degree of freedom to whoever is bound by it, the cord provides greater latitude than the arm or the stick, which can bring about only rigid relations. Like a goat that can graze the circle around its post, within the circumference traced by its tether or halter, anyone attached by a cord can move about with free hands and elbow room within a short radius, and will be held fast only at the limit of tautness.

Law marks limits. The bond makes it possible to feel these borders, but only when it becomes taut, straight: that is, when it becomes law. Prior to that it defines a space, plane, or volume, free and unbound. Or a zone of non-law within law.

Thus the variation before the frontier is reached is just as important as the border itself. If the cord gets hard and stiff, then it imitates solids; at rest, soft, coiled, folded, sleeping, lying looped on the deck, it becomes invaginated, absent. A strange metamorphosis, a natural and scientific change! Think of it as some variable liquid whose density ranges from highly volatile to thickly and intractably viscous: you can be flying and swimming as free as you please, but suddenly the ice takes and you're stuck in its grip. Bound hand and foot, obligated. Cords, moreover, form the main component of your clothing: your well-being inhabits a large coat, which suddenly straps you in. Limits invert the properties they enclose and protect: mobility inside and fixity at the frontiers, absence inside, sudden presence on the borders. Liquid ripples in the wind, clothes billow, a cord makes folds, loops, and braids, but crystallization confines like a straightjacket, and bonds become tight and rigid. Law's straightness surrounds and organizes the looped and folded spaces of non-law.

The technical description of bonds and their knots enables us to bring together and grasp both continuous space and its catastrophic limit, flexible-string topology and stiff-cord geometry (geometry alone can measure, partition, distribute, and attribute), variation and invariance—and thereby combine freedom and constraints. It's as if we were watching science, technology, and law being born simultaneously.

I also like to say that bonds *comprehend,* since they join or grasp or seize several things, beasts, or men together. The bond is doubtless the first quasi-object suited to making our relations visible and concrete; the real chains of obligation, which are light and unburdensome within a space, weigh us down at its edges.

Does a contract mean that together we are pulling or drawing, gripped and yoked to the same trait or draft, the way two oxen, bound together, used to draw or tract the plow? This cord attaches us to other men and to the thing pulled along. The slightest movement within either person's space of freedom can have immediate effects on the thing, within the limits of its constraints, and the thing's reaction in turn acts unopposed on us. This is a system of relations, a set of exchanges. Thus each element of the group, bound, turns out to comprehend—mechanically, by force and motion, in real time—the site of the others, because it is constantly informed by them.

A contract, therefore, doesn't necessarily presuppose language: a set of cords can be enough. They themselves comprehend without words. Etymologically and in the nature of things, a contract *com*-prehends. We are apprehended together and we apprehend one another, intercorded, even when mute; better yet, the contract blends our constraints and our freedoms. The information that we each receive through our tip of the cord informs us, in the end, not only about everyone else who's encorded, but about the overall state of the whole system to which we belong. The bond runs from place to place but also, at every point, expresses the totality of sites. It goes, to be sure, from the local to the local, but above all from the local to the global and from the global to the local. The contract thus affects us as individuals by making us immediate participants in our entire community. It blends solitaries into a collectivity.

This cord has three functions. First, the cord of the *harpedonaptes* marks out a field and with its flexibility surrounds it: can anything be defined without it? To this object, second, it attaches the subject, as if to its knowledge or to its property. And third, it informs others, contractually, of the situation produced by the enclosure: can there be collective forms of behavior without this? These practices concern, respectively, form, energy, and information; they are, if you will, conceptual, material, and judicial; geometric, physical, and legal. Bonds of knowledge, of power, and of complexity. All in all, its triple tress links me to forms, to things, and to others, and thus initiates me into abstraction, the world, and society. Through its channel pass information, forces, and laws. In a cord can be found all the objective and collective attributes of Hermes.

When flexible, it embraces topology, only to describe geometrical forms once it stiffens. By means of brief little pulls, it conveys information, at low energy levels, whereas, when continuously pulled taut, it transmits force or power, high energy levels. At its constraining limits it imprisons, but it leaves elbow room prior to this maximum. In the cord, we find the sciences of space and the genesis of their objects, as well as the technologies of force or energy: who can be surprised that the cord also binds together rigorous knowledge and law?

Moreover the word *trait,* in French, like *draft* in English, means both the material bond and the basic stroke of writing: dot and long mark, a binary alphabet. A written contract obligates and ties those who write their name, or an *X,* below its clauses. In the absence of concrete bonds, hemp yarn or iron chains, and of tightened knots, a treaty, once drafted, remains effective and functions by itself, by the fidelity of a word given or the solemn pact before a notary. We are apprehended by the contract, which in turn comprehends us: we inhabit its network, local and global, held by its system and by all the partners who have countersigned it. It can be easier to get out of a harness than out of a penstroke.

Now the first great scientific system, Newton's, is linked together by attraction: there's the same word again, the same trait, the same notion. The great planetary bodies grasp or comprehend one another and are bound by a law, to be sure, but a law that is the spitting image of a contract, in the primary meaning of a set of cords. The slightest movement of any one planet has immediate effects on all the others, whose reactions act unhindered on the first. Through this set of constraints, the Earth comprehends, in a way, the point of view of the other bodies since it must reverberate with the events of the whole system. This, then, is a contract of universal affiliation. Newton himself would not have repudiated this approach, which reprises that of Lucretius: natural laws federate things just as social rules bind human beings.

When our tools were local, and kept us working on only our own hayfield, we weren't constantly informed of the Earth's global changes. A little harness was enough for us, so that along with a few neighbors we could just manage to draw or tract a narrow plow. The only interesting information had to do with the plot of ground. In those days, outside the field and the village we knew only of the desert and of vague peoples. Our social contract comprehended but a few objects, drawn by a modest number of members. There were always more mouths than bread, and therefore more words than things, more politics or sociology than objects of consumption. There was no nature, in the global sense of the word; the so-called modern social contract is unaware of nature, since for it the collectivity lives only in its history, and that history lives nowhere.

I remember, since I was born there and absorbed its culture, this former world without world, where we were only locally bound, with no responsibility beyond our narrow borders. Whence the foreign and world wars whose untrammeled ravages and atrocities made of us a generation of world citizens.

Today the global power of our new tools is giving us the Earth as a partner, one whom we ceaselessly inform with our movements and energies, and who, in return, informs us of its global change by the same means. Once again we have no need of language for this contract to function, like a play of forces. Our technologies make up a system of cords or traits, of exchanges of power and information, which goes from the local to the global, and the Earth answers us, from the global to the local. I am simply describing these cords so as to speak, in several voices, of science, technology, and law.

Once the angelic bearer of personal messages, the god Hermes used to cross formless regions to speed from singularity to singularity; then an annunciation—the Angelus—was quite an event. But from now on, the name Hermes refers to the totality of bonds of every kind that attach all of humanity to the world's globe, and vice versa. Communication functions are becoming integrated, and, in so doing, are speeding toward a kind of metastability. The progressive erasure of local events constitutes the greatest contemporary global event.

Bound together by the most powerful web of communications lines we have ever spun, we comprehend the Earth and it comprehends us, not just on the level of philosophic speculation, which wouldn't have been all that important, but in an enormous play of energies that could become deadly to those who inhabit this contract.

We've been living contractually with the Earth for only a little while. As if we were becoming its sun or its satellite, as if it were becoming our satellite or our sun. We draw each other, we hold each other tight. In arm wrestling, with an umbilical cord, in the sexual bond? All that and more. The cords that tie us together form, in all, a third kind of world: they are nutritive, material, scientific and technological, informational, aesthetic, religious. Equipotent to the Earth, we have become its biplanet, and it is likewise becoming our biplanet, both bound by an entire planet of relations. A new revolution, in the Copernican sense, for our grandeur and our responsibilities. The natural contract resembles a marriage contract, for worse and for better.

Analysis must be understood as the set of acts and thoughts that unbind. All along every bond, wherever it passes or grips, it transmits force or information, some kind of reverberation. Modern science cut up these bonds so as to institute precision and exactitude, and, by means of these partitions, it refused universal reverberation; the scientific ideal reversed the contract's function. The global problems posed by the sciences and by contemporary needs are in turn overthrowing this ideal of division and cutting, so that they are retying bonds that analysis had untied. We are returning to the contract.

Until this very morning nature eluded us: either we limited it to the local experience of the little hayfield, or else we made it an abstract concept, sometimes applied to man. And if we studied it, in the sciences, we cut it up into even smaller plots; one of the crises in our knowledge comes from its inability to function without these divisions and from the need to solve the problems posed by their integration. Here, then, is nature today, new and fresh, being born: global, whole, and historiated before the eyes of global humanity as a whole; theoretical, soon, provided that the disciplines are willing to join in federation; concrete and technological right now, since our means of intervention act on it and it in turn acts on us; a network of multiple bonds where all things, congruent, conspire and consent; a web tied, by a lattice of relations, to the henceforth united social and human fabric.

The sum of these cords, stitches, and knots, assembled in various latticeworks, interconnected throughout, defines nature in a simple, clear, distinct, speculative, and technical manner, in a way that was sometimes dreamed of in the past but that was certainly never conceived or put into practice. Nature is a set of contracts.

Curiously, it is only in this century that nature has been born, really, before our eyes, at the same time as a humanity bound in real solidarity, not the verbal solidarity of official speeches. The silhouette of the great Pan, the demon of globality, appears at last behind his father, Hermes, god of bonds. First, his shadow.

Casting Off for the First or Last Time?

Go look for death where it is prowling, free and naked, active and scattered in all directions, attentive, never sated, and you will discover the other world, organized and defined by its constant diligence. Down in this world you allow yourself a thousand peaceful acts: to sleep, dream, talk on and on, relax your attention; all danger moves away from your steps so naturally that you don't think about it. Houses and gardens, hedges, plowed fields, stores, schools: everything is sleeping or purring; here death strikes rarely, as if it came from elsewhere, and when it does everyone is astonished. But in the world beyond, it reigns in every detail and space, like a dense presence.

Beyond the port: shipwreck for the smallest error; once past the mountain hut: at the slightest mistake you will fall; cast off from the launching pad: at the first moment of inattention the rocket will explode, killing the seven crew members; for a minimal aberration an accident results. Minor causes, great effects. In one's bedchamber, everything is forgiving, the bed and the pillow, the armchair and the rug, supple and soft. A thousand causes with nonexistent effects.

Walls, cities, and ports, havens from which death keeps its distance.

Beyond, death roves through space, prowling. Never sated, it nests in low, black caverns; everywhere it lies in wait and yawns. Once you cast off, everything you do can be held against you. The words of the examining magistrate resound. High place: high court. Here the causal space of cases is open, with no apologies or forgiveness. Every act counts, every word and even intention, down to the slightest detail. Like a judicial proclamation, an act accomplished here is immediately performative. Reality clings to it: no sooner is an act begun than it is subject to sanction. You no longer have the right to fall. You begin to live in another way. Neither bed, nor wall, nor hedge keeps you from death.

How to define our ordinary world? "That doesn't count": this is the only rule or, better, the gap in its laws, the cord's braids and loops. A thousand things without importance are neither obligatory nor punished here. You do not have to pay for every detail of common life. A hundred spaces beyond the law let you do, say, or get through as you wish. Customarily, non-law prevails over law. The ease of our bodies comes from this elbow room. Who would complain about these degrees of freedom, this gratuitousness that makes up life itself? Here the cords of contracts go slack, over there they are taut.

No other world is forgiving: death is watching and punishes all sorts of mistakes. Whence the demand for constant control, which teaches virtuosity by force. The guide comes away from the wall of vertical ice without making a mistake: in other words, he doesn't die, he maneuvered in a causal space where everything counts, and there he practiced virtue, which must be defined as that which permits virtuosity. Because he's at the head of the roped party, he's unbound: only the connection or the contract gives security, along with obligation. Assurance comes only from competence, for anyone who finds himself alone, having no bonds except with the thing itself. With the sheer wall of the world.

The transcendent virtuoso passage on the violin or the piano is executed with precision by the master, whereas another player would get hung up on it. Of course no one has ever died from playing a wrong note; nonetheless the whole career of any virtuoso is decided at each instant in such passages. He is not playing an instrument, but always playing and betting his whole existence. If the taut string catches, the partita rings false; elsewhere, in science, for example, losing an edge makes measurement fail and truth vanish; let the smallest vowel get out of place and see how the page grows ugly, dismaying. Proofs, the sea, great art, and ice do not allow false steps. Beauty never enjoys the right to err. At the first sin, hell opens its maw.

Sanction and sanctification, both derived from the sacred, which is produced by death, go back to the same origin: other worlds prove, willy nilly, to be sanctioned spaces, places of law and causality, sanctified places; such worlds are the house of the solitaries, hermits or anchorites, immersed in the worldwide world.

Higher mathematics, fine arts, high virtuosity, high-level competition, high mysticism, all correspond in every way to high mountains or the high seas, worlds where the cords remain taut.

Abstract or concrete: the most concrete, sea and mountain, seem abstract to some, and the most abstract, algebra or musical notation, seem concrete to others; those worlds surround this one, just as long ago, before Christopher Columbus, model of virtuosi, unknown continents bordered the places thought to be the only ones inhabited. Strangenesses surround our space. Casting off leads us to them.

Our tranquillity chases death off to these neighboring and remote worlds, to these third worlds. Everyone considers these worlds dangerous, but what they actually call for is simply presence, because there one must respond, at every point and in real time, to the active attention of death; one must be as alert and present as it is, in order to reply to it tit for tat. Granted, death does not actually attack—aggression must not be its character—but it is passive like a black hole, it takes everything that is neglected and punishes without fail. That makes you very supple, very intelligent; that keeps you awake. Diligence against negligence.

In this world, everything sleeps. In the other worlds, all the solitaries keep a wakeful watch. Where can one breathe air more alive? The sleepers join together in the common world. Elsewhere the wide-awake disperse.

Thus when I think, I really only think in and through one of these other worlds, where only forms of vigilance can dwell, get through, or even exist. Truth, thought, meaning, even awakening must be won from death, for nothing is better than death at invading a space completely and obliging you, at every step, to be a virtuoso. An instinctive instigator, an instructor, death alone, like hunger, teaches us what we must know. The rest does not even deserve the name knowledge. The verb *to educate* means indeed to lead elsewhere, out of doors, outside of this world: in fact, to cast off.

Here, I doze off, in this world I rest. Here lies.

So all my stories and the whole universe are reversed: assurance puts us to sleep, ordinary life gives itself over to death, the death in which normal stupidity, repetitive and limited, slumbers, drugged and bound—whereas the other worlds are populated with the lively and hardy. The taut. All in all, no more of them die than of the sleepers. Death vivifies life, which dies from lack of death. Depart—toward nature—to be born.

Sown everywhere, behind each rock, under the crest of every wave, ready to nip at your buttocks, death is a more than perfect training, driving you to continuously excellent action. Never leaving death's implacable school, the strong and healthy life is dedicated to work. This is the secret of all production, this is why culture only finds refuge in third worlds. Life that is good is interested only in death, which, in exchange, shapes it.

Once past the other worlds that stimulate our own, we will cast off anew toward death, our origin. To be reborn.

Palo Alto, after October 17, 1989, at 5:04 P.M.

For the past twelve to fifteen nights here, everyone, in the privacy of his bedroom, has set at the bedside, just before turning in, a sweater, a flashlight, and a pair of slippers, emergency equipment in case of a strong earthquake. Scientists and experts advise preparedness.

And so, every evening, I lay out and look at the wretched little pile of rags on the floor, the bare essentials one can grab up in a second, ready for departure, and I play out this scene in my head: get up hastily, remain calm, put on slippers, quickly turn on the flashlight . . .

. . . but why, to go where, and especially when, at what time, for a shock of what intensity? Admittedly, the Earth has not stopped shaking here for more than two weeks, but as far as I know anything can happen at any time, and not just after an earthquake. Must one always cling to an emergency kit, a viaticum?

What strong and simple science will tell me the moment of denouement, of being stripped bare and untied, the moment of true casting off, and tell me to take nothing, to go completely naked, overwhelmed, burning, trembling in all my limbs, from this Earth toward the void, or toward what tremendous god of love?

Anne, Mother of Mary

Hard and generous, stiff, surly, completely obliging, muscular and brash in the country way, poor, never married, the eldest had never left the town nor the house of her parents, which she had ruled over rigidly and unfailingly ever since her mother, whose empire had lasted a half-century, began to give way. As far as anyone knew, she had had no affairs, and had no particular faults or notable talents, no emotional life. Up to the age of sixty or more, her life flowed by, compact and inflexible, without anyone seeing her eyes mist over with tears. A certain kind of religious and moral education suppresses the person, for better and for worse.

Near the Christmas holidays that year, her mother, whose mind had long since gone back to the joys and naïvetés of paradise, took to her bed to set about dying. One of those powerful characters who never take naps and whose first rest coincides with their last, she spent an interminable time passing away. Luck would have it that, of her eight children, the five girls were attending her piously in the simple and solemn moments when life hesitates to rise in the air and leave behind the reluctant mortal remains lying there.

Did a sudden shock hit her? Shattered by this blow, the eldest got up, took her mother in her arms, and began to walk around the room, carefully, with rhythmic steps, singing a childhood melody whose monotonous chant drowned out both the hymn sung by the sisters, on their knees praying, and the death rattle.

She was rocking her mother's pitiful body this way, against her stomach and in the cradle of her elbows, when those watching saw her face, close to the mouth breathing its last, transfigured: gentle, very sorrowful, radiant with goodness, tranquil, sublime. . . . She was giving birth, she was opening her mother's way to another life, by birth or by resurrection, and she was accompanying her patiently in this supreme effort like the woman who gasps and pushes in labor but seeks to reduce the violence and the forcing in order to spare the body of her little child.

Then the crazy grandmother died in the bosom of her sterile daughter, through supernatural maternity, while the death rattle and the lullaby mingled with the hymn of the four other daughters, natural mothers, whose toneless voices followed these two mysterious passages, mystically joined.

Without words, amid linen, scattered towels and hanging sheets, unfolded wet handkerchiefs, dirty cloths, all of this happened right at the level of life and death, led by the body, biologically, savagely, archaically, doubtless by repetition of what unimaginable ancestors must always have done without knowing why, or simply because having two feet, two fists, a sex, and a head, they were letting themselves flow along in the hominid lineage through the feminine channel, through the birth canal.

The mother goes back or leaves through the belly of her virgin daughter: Anne Mary.

In a remote Chinese forest, six or eight woodcutters are preparing to lift up a monstrous log, from one of those trees with wood as dense as tempered steel, a trunk lying there on the ground, stripped and lifeless, whose diameter by far exceeds their height. They will never be able to carry this gigantic mass. Gently they approach it, as if to tame it; they touch it in certain places that they seem to recognize, examine it in silence, very slowly tie it with simple cords, and cover their shoulders with old folded sacks wider than their loose, tattered loincloths.

They are almost naked, I remember now, and their hair is graying, their pointed beards whitening. Ceremonial movements delicately close to the wood, and harmonious manners infinitely close to one another. Now they are bent over, the lines seem to become taut; the log does not budge.

Then a sheet of sound, a song, bathes the whole scene; it comes from who knows where, from the forest, perhaps, the copse, the surrounding foliage. Barely sonorous, husky, low, sweet, it emanates, still submerged in the entrails: is it possible that a noise can partake less of the audible than of the intimacy of the living bodies nearby? That a sound can remain still immersed in the mass? The prostrate backs were singing, praying, groaning, seeming to curl up in a childhood lullaby; they were calling the beam, or madrier, which was responding to them with some mystical Gregorian chant. The trunk put out new roots in their thighs or came out of their loins.

I am telling you what I saw and heard: matter raised itself. Yes, transported by the seven stocky foresters in the cradle of liana, which vibrated like the strings in the low notes of the piano. But no. Matter levitated. Borne up by the breeze of the music, the beam took to sail; it cast off.

I am relating here a very ancient account: I do believe that in our ancestral languages, the terms *madrier* and *matter* meant both wood and mother.

But the word cometh always: at the very moment when the grandmother was dying, delivered by her virgin daughter, borne off by the mad wind of the hymns, the door that no one had thought to guard opened violently, pushed by some rushing mighty tempest, and the eldest of the great-granddaughters, age seven, redheaded, rugged, and hardy, came in, full of fire, brash and muscular, holding in her hand a sheet all scribbled over: "Here," she cried, "here's the letter I've just written for grandmother who couldn't anymore. It has to be put in the coffin so the Good Lord can read it when she arrives."

Words and flesh, the corpse cast off equipped with the program.

Sequel beyond the Grave

Psychopompus: this is one of the names under which Antiquity venerated Hermes; by this title they meant that he accompanied dead souls to hell. He watched in silence over our mortal agonies, guide of messengers, bonds and cords, angel flying in limpid air, nimble as a rocket, then led us toward the other world. His name, his acts, his myth sum up all these stories.

He was honored, as well, as an innovator: he had invented objects, the lyre and the panpipe, named for his son, but also the letters and signs of writing; and perhaps, too, road milestones, those tall rocks that in ancient Greece bore his name as well as a face and a sex, organs of communication that symbolize roads.

Constructor of relations, of objects, conductor after death, god of messages and productive passages, his silent and translucent presence could be divined at the two twilights of dawn and dusk. In sum, Hermes could have passed for the archangel of casting off.

The lovers' apple, a token exchanged between our first parents, weaves solid or fragile bonds that adversity often severs; connections build the ship, and the shuttle-fruit traces relations in spider's threads; communications techniques produce *Ariane,* which multiplies them and magnifies them into telecommunications satellites. In general, relations, sometimes de jure, construct objects, always de facto, which permit relations, which, in turn, produce other objects: we inhabit this spiral curve, continuous, broken or turbulent.

What could be more obvious than the god of messages and interpreters becoming a skilled artisan? Did he make the first cords? Through identical sounds, our language says: translators and conductors quickly become fabricators and constructors. The guide hidden behind these rhymes or roots hands me a bond, a fabricated object, a reliable relation, then a contract.

Casting off throws us elsewhere, or toward and into another world, so that this relation causes a craft or piece of gear, an "object," to appear: in the literal sense, a thing cast before us. Of course it must have left our bodies, to be lying before us like that! Where would it come from, otherwise, this casting that bursts forth and takes off? Sometimes the whole organism hurls itself outward, the functions of its organs casting off to become tools. The projection comes from the subject, once again well named. In contrast to animals, enclosed in the stable armor of their instincts, let us call man this animal whose body leaks, its organs becoming objects.

So behind these symbols, these people and their acts, is hiding the one from whom the guide protects us and who educates our steps, who leads us and at the same time obliges us to produce: death. Our castings off toward it force us to make tools that have cast off from things, words that have cast off from artifacts, music that has cast off from words, mathematical signs that have cast off from music. . . . Departure toward death informs and sums up all the other departures.

Example: after thousands and thousands of millennia of infernal efforts, Sisyphus succeeds in pushing the tombstone out of the earth; the dead body of Hestia, funereal goddess, appears on our road, a cairn, bust, or pile of rocks, a tomb; the latter becomes one day an enormous pyramid, a statue or a colossus, a tower; later, carved, full of holes, delicately wrought, as if emaciated, tremendously animated, a sort of Eiffel Tower casts off, in the hurricane and clouds—this is the rocket amid the stars. Thus the death of Hestia, virgin and mother, catches up with the launching of *Ariane,* blindingly shorting the endless and patient circuit of hominization. Our most sophisticated objects result from a succession of castings off, with death always right behind. This long and true story develops the most archaic chapter of the overall epic of the god Hermes; the following chapters, the easiest ones, sing of music, speak of language, and decipher the sciences, finally reaching our present-day achievements.

Earth Ho!

But adversity, which sometimes severs bonds, is no longer attacking just our body, already destined for death since the dawn of time, and defending against death by precisely this way outside itself or these connections—apple, cord, or masterpiece, the page written in a pathetic or banal tone. Adversity is now attacking that which attaches and binds us all together and connects us universally, our earth and our species, complete sums of our cords and alliances. Since Nagasaki we have our disappearance in our power, and the danger curve is rising exponentially. Though I have been deaf ever since the dominators of this world began thundering shamelessly, I hear, and I'm not the only one, the revealing hissing of air strata driven down by enormous falling rocks. Individual, local, ancient, and primitive death is being succeeded by a modern, specific, and global death, our collective worldwide horizon.

Is this modern death going to awaken us from scientific sleep, and for what other casting off toward what excellence or virtuosity? Will it give back to us as much intelligence as the inventors of the sciences received in the past from its archaic sister? The more pregnant the death, the more capable our efforts, the greater the scope of our object-worlds.

To universal death corresponds, then, to push the point, the universe as object. Cast before us, here is Earth. Does she come out of us, do we come out of her?

From the nature we used to speak about, an archaic world in which our lives were plunged, modernity casts off, in its growing movement of derealization. Having become abstract and inexperienced, developed humanity takes off toward signs, frequents images and codes, and, flying in their midst, no longer has any relation, in cities, either to life or to the things of the world. Lolling about in the soft, humanity has lost the hard. Gadabout and garrulous, informed. We are no longer *there.* We wander, outside all places.

Cast off far enough from our Earth, we can finally look at her whole. The farmer with his bent back lived on the furrow and saw nothing else; the savage saw only his clearing or the trails across the forest range; the mountain dweller, his valley, visible from the mountain pastures; the city dweller, the public square, observed from his floor of the building; the airplane pilot, a portion of the Atlantic. . . . Here is a hazy ball surrounded by turbulence: Planet Earth as satellites photograph her. Whole.

How far up must we be flying to perceive her thus, globally? We have all become astronauts, completely deterritorialized: not as in the past a foreigner could be when abroad, but with respect to the Earth of all humankind.

In bygone days, each individual, at once a soldier and a tiller of the soil, used to defend his plot of earth, because he lived off it and because there his ancestors lay: the plow and the gun had the same local meaning as the tomb, object-bonds to the soil. Philosophy invents the being-there, the here-lies, at the very moment when this localness is disappearing, when the earth comes together and moves from plot of ground to universe, when its name is adorned with a capital letter. From this small local port and its ordinary objects we have cast off. Our most recent voyage brought us from earth to Earth.

All humanity is flying like spacewalking astronauts: outside their capsule, but tethered to it by every available network, by the sum of our know-how and of everyone's money, work, and capacities, so that these astronauts represent the current highly developed human condition.

Seen from above, from this new high place, Earth contains all our ancestors, indistinguishably mingled: the universal tomb of universal history. What funeral service do all these vapor plumes herald? And since, from up here, no one perceives borders, which are abstract in any case, we can speak for the first time of Adam and Eve, our first common parents, and thus of brotherhood. One humanity at last.

Here is where our expulsion from earthly paradise led us: this, then, is the provisionally final result of hominization and history, of our work, of the painful generations drawn forth by individual death. To the universe-object corresponds, then, in all its meanings, universal death: of course it threatens us here, but it is also lying in wait everywhere else; what I called the other world now covers the whole planet.

For the first time, philosophy can say man is transcendent: before his eyes, the whole world is objectifying itself, thrown before him, object, bond, gear, or craft; man, for his part, finds himself thrown outside, totally cast off from the globe; not from the port of Brest, the Kourou base, the mountain hut of Chabournéou, or from his death bed, not from a given place, here or there, not from the humus of his life, his paradise, no longer from his mother's entrails, but from the whole Earth....

...The largest apple. The most beautiful sphere or turbulent ball. The most ravishing boat, our caravel new and eternal. The fastest shuttle. The most gigantic rocket. The greatest space ship. The densest forest. The most enormous rock. The most comfortable refuge. The most mobile statue. The complete clod of earth open at our feet, steaming.

Indescribable emotion: mother, my faithful mother, our mother who has been a cenobite for as long as the world has existed, the heaviest, the most fecund, the holiest of maternal dwellings, chaste because always alone, and always pregnant, virgin and mother of all living things, better than alive, irreproducible universal womb of all possible life, mirror of ice floes, seat of snows, vessel of the seas, rose of the winds, tower of ivory, house of gold, Ark of the Covenant, gate of heaven, health, refuge, queen surrounded by clouds, who will be able to move her, who will be able to take her in their arms, who will protect her, if she risks dying and when she begins her mortal agony? Is it true that she is moved? What have we not destroyed with our scientific virtuosity?

Emotion: that which sets in motion. How will we move, on the day when we no longer lean on her? How can we rock her in our arms without anchoring our feet on her stay? Or cast off from her without her? How, then, to be moved? Those who lose the Earth will never again be able to cry. They will never again be able to bury their ancestors. We never weep for anything but the loss of a mother, she who rocked us in her arms, the only consoler of all our afflictions. Heroic, surely; intelligent, of course; brilliant, why not; but inconsolable and unconsoled.

Flying high enough to see her whole, we find ourselves tethered to her by the totality of our knowledge, the sum of our technologies, the collection of our communications; by torrents of signals, by the complete set of imaginable umbilical cords, living and artificial, visible and invisible, concrete or purely formal. By casting off from her from so far, we pull on these cords to the point that we comprehend them all.

Astronaut humanity is floating in space like a fetus in amniotic fluid, tied to the placenta of Mother-Earth by all the nutritive passages.

From the highest place we have ever reached, in all the castings off of history, the universal-subject, humanity, in solidarity at last, is contemplating the object-universe, Earth; but also, the infant suckles its mother, still attached to her by so many cords and threads. Thus the bonds of life or of food and those of thought or of objectification become alike in emotion.

From this site, our here and now, the new place of our contemporary existence and knowledges, from this place, whence philosophy now observes and thinks, technology is becoming more like organic life and science more like nature, in the sense in which nature means an upcoming birth. Through the canals and channels of these multiple bonds, hard and soft, who will give life or death to whom?

For this new subject bound to the new object, life and death trade roles once again, dangerously, to reach an even higher level of virtuosity. Must we not become, in effect, the mother of our old dying mother? What unheard of meaning would this new obligation have: to give birth again to the nature that gave birth to us? Is Earth a Virgin who gave birth to her Creator, Him or Her?

Yes, Earth is floating in space like a fetus in amniotic fluid, tied to the placenta of Mother-Science by all the nutritive passages.

Who will give birth to whom and for what future?

Casting off or parturition, production or childbirth, life and thought reconciled, conception in both cases: would great Pan, son of Hermes, return in this mortal danger?

These symbiotic bonds, so reciprocal that we cannot decide in what direction birth goes, define the natural contract.

In Distress,

this is my signature; for, most often, I live and feel myself in distress, just as amid the fearsome hurricane and sea, a vessel quickly loses all its gear; the waves ravage the tops, the masts break, the network of ropes tears, everything goes into the water, and all that remains is the dancing hull, full of holes, which the surviving crew clings to. I have been surviving in distress for so long that I have lost all deadwork of my own, flag and title, ties, sails, coat, address and port, name, face, bearing, and opinion.

To cast off means that the boat and its sailors entrust themselves to their technologies and their social contract, for they leave the port fully armed, from head to toe, with proud yards and boom aimed toward the future. It appears that, by taking to the water, they are taking the sea in their craft's gear and tackle: the ship inhabits its ropes and whaleboats, surrounded by its prow and rudder, protected in the cage of its knotted ropes; the helmsman inhabits the boat. But all these fine people, so well prepared that they announce, at the start, that steps have been taken against every risk—they cast off a second time when the storm rips out cables and capstans and undresses the skiff by tearing the fabric of its rigging: henceforth disabled and in distress.

I do not want to remember the days when I crossed this second stage, essential and true; since then, I have no more gear on my craft—it even seems to me that I never had any. Since the turmoil of my youth I have gone naked. Reduced to the bare leftovers. I am even missing much of the indispensable baggage for living comfortably. I live in shipwreck alert. Always in dire straits, untied, lying to, ready to founder.

Does a beautiful and good life devoted to powerful works require these irremediable losses? Do serenity and great health positively love the terrible bloodlettings that come with endings and undoings?

That's why I tasted joy during the earthquake that terrified so many people around me. All of a sudden the ground shakes off its gear: walls tremble, ready to collapse, roofs buckle, people fall, communications are interrupted, noise keeps you from hearing each other, the thin technological film tears, squealing and snapping like metal or crystal; the world, finally, comes to me, resembles me, all in distress. A thousand useless ties come undone, liquidated, while out of the shadows beneath unbalanced feet rises essential being, background noise, the rumbling world: the hull, the beam, the keel, the powerful skeleton, the pure quickwork, that which I have always clung to. I return to my familiar universe, my trembling space, the ordinary nudities, my essence, precisely to ecstacy.

Who am I? A tremor of nothingness, living in a permanent earthquake. Yet for a moment of profound happiness, the spasmodic Earth comes to unite herself with my shaky body. Who am I, now, for several seconds? Earth herself. Both communing, in love she and I, doubly in distress, throbbing together, joined in a single aura.

I saw her formerly with my eyes and my understanding; at last, through my belly and my feet, through my sex I am her. Can I say that I know her?

Would I acknowledge her as my mother, my daughter, and my lover together?

Should I let her sign?

French philosopher of science **Michel Serres** has taught at Clermont-Ferrand, the University of Paris VIII (Vincennes), and the Sorbonne. He was elected to the *Académie Française* in 1990. He has served as visiting professor at Johns Hopkins University and has been a faculty member of Stanford University since 1984. His other works available in English translation include *Conversations on Science, Culture, and Time* (with Bruno Latour); *Genesis*; *Rome: The Book of Foundations*; *Detachment*; *The Parasite*; and *Hermes: Literature, Science, Philosophy*. Forthcoming from the University of Michigan Press are translations of his works *Le Tiers-Instruit* and *Statues*. **Elizabeth MacArthur** is Associate Professor of French, University of California, Santa Barbara. **William Paulson** is Associate Professor of French, University of Michigan.